OUT

OF

ASIA

OUT
OF
ASIA

JUSTIN THYME

authorHOUSE®

AuthorHouse™ UK
1663 Liberty Drive
Bloomington, IN 47403 USA
www.authorhouse.co.uk
Phone: 0800.197.4150

Published by AuthorHouse 01/23/2015

ISBN: 978-1-5049-3480-0 (sc)
ISBN: 978-1-5049-3479-4 (hc)
ISBN: 978-1-5049-3481-7 (e)

Contents

Preface

The contents of this book is about an alternative point of view about human evolution of which you may not have heard elsewhere. Also it tries to explain why the races of mankind are different in the way that they are, also the reasons for racism and why it is so prevalent amongst those that live the closest to the minority group. It will also try to explain why countries with a certain racial balance do not do so well economical, or have the same economic equality of other countries that have a different racial balance. This may have nothing to do with intelligence.

This book does not set out to justify racism in any victimising sense, but only to explain why social preferences may exist.

The political correct will always give explanations that demean the "racist", since this is the easy way out and fits in with their left wing views.

Understanding the Anthropologist

I would point out first that professional anthropologists rely on university grants to fund their work, as anthropology has little value in the commercial world. This means that the universities are reliant upon government funds. Even during the best of times, an anthropologist may find his or her grant curtailed because the university finds other priorities to fund. However, any anthropologist making waves and becoming too controversial finds him- or herself in the vulnerable position of losing grant monies and finding the rug being pulled from under his or her feet. A university's excuse for withdrawing funding is sometimes expenditure problems, i.e. the government pulls the strings and determines where the money goes. Because of this, one must not be surprised by professional anthropologists shouting, "No proof!" after hearing any controversial statement.

NB: Absence of Evidence Is Not Evidence of Absence

There are two types of anthropologists: the "stone and bone" anthropologist and the "hunter-gatherer" anthropologist. Stone-and-bone anthropologists appear to love the dramatic. One of this type is Robert Ardrey, a "writer" and "dramatist" who believes that humans are killer apes. Ardrey sees a human being on a hilltop, pointing his or her weapon to the sky, as the "lord of creation," which is how we sometimes see ourselves – that is, until some natural catastrophe puts us in our place. I must confess that I, like Robert Ardrey, do like the dramatic. Therefore, I warn the reader that I may have a bias for the more carnivorous way.

The hunter-gatherer anthropologists see a human being as a kindly, happy-go-lucky ape wandering around the earth eating mangoes and nuts, hunting for the fun of it, and keeping busy while evolving a large brain by sitting around camp fires talking about and contemplating their nuts. It is, generally speaking, people on the left wing of politics (or people who support women's liberation) who tend to believe that a human being is a more "passive" hunter and gatherer. Espousing this image of humankind implies that hunting is less important than food gathering. Also, since hunting is something that women find difficult to do (although not, I would say, impossible, since some Australian women are known to do some hunting, e.g. fishing with spears. This, however, is not very dangerous and hardly requires high vascular activity), the conclusion can be reached that men are not very important and that women could do without them. Some primitive cultures retain a hunting-and-gathering way of life. I wonder why they are still primitive cultures, ones that remain technologically behind European Neolithic man, as he was twenty thousand years ago. I personally believe that humankind evolved as killer apes but that there were offshoot tribes or clans that may have been vegetarians. Anthropologists may disagree on this point. Those who lean to the left believe that humankind has evolved to be good, kind, and nice. They also hold that we are all born equal and that our differences developed in light of environmental or educational circumstances. (Perhaps humankind evolved into its present "perfect" state, in God's perfect image, of course, and then God put the mockers into our gene pool to stop us from evolving [or devolving] any further.)

Right-leaning anthropologists believe that human beings have evolved to be warriors, hunters, and killers. They espouse the idea that people are the way they are because they were born to be that way – and one cannot make a silk purse out of a sow's ear. Perhaps the whole truth is somewhere in between. Today's prevalent opinion, one held especially by left-wing, or "politically correct," anthropologists, is that people are born with an innate goodness but have fallen from grace. In other words, a human being is a fallen angel. The opposing argument maintains that a person is born a hunter, a killer ape – a creature primed for war and conflict. In other words, a human being is a risen killer ape. The normal left-wing response to this assertion is, "You don't think very highly of yourself, do you, when you say that you are no more than a killer ape?".

If human beings are born angels, then what are our achievements? Music, compassion, peace treaties? Well, no, those things are not achievements for an angel, any more than feathers are an achievement for a bird, since there is no free choice involved. It is only what one would expect of an angel, one who is born with such goodness. Humankind's only achievements are waging wars, committing atrocities, sending others to the gas chamber, and committing inhumanities against other human beings.

If humankind consists of risen killer apes, what, then, are its achievements? If you say wars and atrocities, then you point to what can only be expected of a killer ape, No, humankind's achievements are music, poems, peace treaties (no matter how many times they may be broken), technology, and his flight to the stars. "I would rather be a risen killer ape that's on the way up than a fallen angel that's on the way down.".

I am not suggesting that people are naturally evil and that we should give up the ghost and cease expecting good, civilised behaviour of one another. On the contrary, I am saying that humankind is more than the sum of its genes. Human beings, I believe, have free choice and may choose options which contradict their impulses, but this does not mean that every individual will automatically do so. We are all vulnerable and make choices in light of the behaviour of others who are part of the society in which we live.

The purpose of this book is not to explain every aspect of human evolution. That said, I do hope to fill in a few gaps and provide information that I have not found written elsewhere. I do not write this book with the intention of carving its ideas in stone. Rather, I simply offer a point of view backed up with a variety of facts, circumstantial evidence, and a bit of guesswork. At times, I use a bit of tongue in cheek. I do not believe that anyone should take him- or herself too seriously. Instead, one should see *facts* in their context and try not to think that all statements are a personal attack upon oneself. After expressing my evolutionary ideas, I will delve into race, economics, politics and philosophy. None of these things is very far removed from the others.

Chapter 1

A Few Notes on General Evolution

Although this book is about the evolution of humankind and the philosophy associated with the theory of evolution, I will cover a few evolutionary problems that are relevant to humankind today. This will become clearer later.

Speciation

The accepted explanation for what drives an individual organism is that it seeks to reproduce its own genes. However, the simple fact that sexuality exists – which, of course, means mixing one's genes with someone else's, after which time the process of natural selection replicates the best genes – is no guarantee that the original genes will be the ones to survive in successive generations. In fact, when you consider the diversity of and change in genes in progressive species, you see that it is anything but the preservation of the original genes. It would appear that to leave descendants, or descendants of one's relations, is the aim. Having said that, there is no aim, as far as evolution is concerned. What survives, survives. That's all there is to it. Genes do not have any conscious thoughts about whether to survive or not. It's the same with animals. No animal has any knowledge of how to reproduce its own kind. I doubt whether any animal knows that intercourse has anything to do with birth or that a male animal knows that his mate's young have anything to do with him. Life has survived because certain genes make the organism behave in a certain way that is conducive to its own survival and reproduction. It has been said that a chicken is an egg's way of making another egg. Whether the animal is aware of this or not, it does not alter the fact that reproduction simply reproduces particular genes in future generations.

I would suggest that no male animal but the human animal wishes to father children. The explanation for a male lion's taking over a pride of lionesses and killing off the previous male lion's cubs in order to bring the lionesses into season is complete rubbish! How does a lion know about reproductive cycles or that the lionesses will be sexually active if he kills the cubs? By experience? A lion has no more knowledge that the future cubs will have anything to do with him than he does of the cubs that existed to begin with. The reason he kills the cubs is because they are good to eat, or because he can't stand the smell of them, or because he simply hates them since they are weak and easy to kill. The real question is, Why does he not kill the cubs that are born months after his arrival? This may be because the cubs smell similar to the lionesses with whom he has become familiar. He may then be triggered with a paternal love for the cubs, but only because he was around at the time of their birth. Finding the exact subjective reason for why the lion behaves in this fashion is somewhat difficult. What I would say with confidence is that any association the lion has with genetics, or "blood," is purely fictitious. When stags defend a stamping ground, it has more to do with territory and the number of hinds that will be attracted to it. The stag, I would suggest, has no concept of fathering future generations. In fact, the stag is more like a latter-day human womaniser who has not the slightest interest in fathering children (and who may go to great lengths to avoid producing a child). The pleasure is in humping as many females as possible. To say that an animal does this in order to do that is only a way for zoologists to simplify and explain the genetic survival outcome of such actions. I will use such terms myself, for the same reason.

It is generally accepted that evolution is based on the principle of survival of the fittest. I question this. Take the case of plankton in the sea, which whales consume by the ton. Does this mean that the fittest plankton are able to swim out of the way? Of course not. The weak and the strong get eaten up just as quickly. There are many cases in nature when an organism is eaten because it does no more than encounter bad luck. In fact, evolution is more a case of survival of the lucky, since, in most cases, dying before reaching full maturity is the norm. The lucky may then compete amongst themselves to find out who is the strongest or the fittest.

Another point I wish to make is that genetic selection is always discussed along with evolution. The two are related indirectly. Strictly speaking, biological selection is the real selective factor involved. When a lion hunts a zebra, the lion does not know anything about the genes of a zebra. It simply goes for the easiest meal it can get, whether or not the Zebras weakness is genetic or environmental. It is only a general fact that this characteristic is reflected in the genes of the Zebra.

On the subject of genomes, if it looks like a duck, flies, swims, walks, and quacks like a duck. Then guess what it is? Well it's a duck. As a philosophical point only. Whether the duck has a "chicken" or "swan" for for an ancestor is really neither here nor there.

It is quite possible that two individuals of the same species who are living in the same environment will produce different genes to adapt to changing conditions, so all the survivors may not necessarily be genetically the same. In other words, there is more than one way to skin the cat. Of course, when two individuals that are adapted to a similar environment because of two different genes happen to interbreed, they both confer upon their offspring their particular good genes. Therefore, the offspring have two differently adapted genes for a similar characteristic instead of one, therefore creating a double whammy or quantum leap in evolution.

Speciation by Migration

When it comes to speciation, which is when a certain pair moves into an area uninhabited by their own species, one might ask, What is the survival incentive, and how is the pair relatively unsuited to that environment? I would suggest that, although the environment may be "unfriendly," the two do not have to compete with their own kind for territory. That uses up a great deal of time and energy, in most cases. Therefore, they only have to survive and make sure that they produce as many young as possible. If they only manage to bring two or three individuals to maturity, then those individuals will be the best from the batch, the ones that are suited to survive the new environment. Also, they don't have to compete with anybody else's young, which, in turn, will carry on those advantages for future generations. Each generation, in turn, adds its own advantages. These species could declare themselves as winners. In some cases, an individual organism may find that a different but similar environment may suit the differences that it inherited. For instance, a more heavy-coated animal may choose to live in a higher region because it feels more comfortable in a cooler climate. So, some organisms may find a new niche by choice, not remain as malleable slaves to their environment. Having fewer individuals of their own species to compete with, they could declare themselves as winners. There are, of course, many instances of successful breeding when the species presents a convincing display of threat to potential contenders. This means that the species does not necessarily exhibit greater strength, fitness, or even intelligence but that it merely gives the impression of possessing these qualities, which allows it to compete successfully within the group. It would be true that such individuals are successful, but a group that concentrates and expands on this advantage may come unstuck if an individual from another territory ignores the threat display. This issue is relevant today, whether a species is competing to overcome the difficult physical problems of living, or even surviving, in the environment or it is competing with others of like kind for limited but easily available resources. I will delve into this later in this book.

The polar bear and the grizzly bear are considered as two different species because they have adapted to two different environments. However, the two bears have been separated only within the last few hundred thousand years, meaning they can still interbreed and have fertile young. Therefore, they could be called the same species. Well, the human races have also adapted to various environments and produced various types of human beings. Does this make them different species?

Dead Ends

One strange thing about evolution is that it is often talked about as if it has a mind of its own, as if evolution did "this" in order to do "that." Unfortunately, evolution is blind and, like all blind entities, has a habit of travelling down dead-end alleyways. I cite as an example something I saw on a TV documentary about a mammoth that has survived into recent times on a small island in Siberia. The question is, Why did the mammoths die out? I would like to contribute an explanation. First, it was a large animal well adapted to the cold that was isolated on a small island. Any species needs a number of individuals to keep the population going, but a small island can only support a finite number of large animals. The mammoths were therefore reduced in size and became a pygmy form. This meant that those individuals whose resources were used to build thicker coats and generate more heat (rather than increase in size) survived better for a time. The population grew to an environmentally affordable rate. This may well have happened in relatively warmer times, too, but if the climate became extremely cold, then the winters would perhaps be impossible for even the hardiest to survive. The problem is that a smaller animal, as opposed to a larger one, has a larger area of skin relative to its bulk. Since an animal loses heat through its skin, it would either have to generate more heat or produce more tissue to provide a thicker form of insulation. Because there is only low-protein food available on the Siberian island, the small animal would have to consume and digest huge amounts of food to produce the thickness of hair and skin it needed, given its body weight. Also, a small animal digests food less efficiently than a large one since the large animal has a large stomach in which to "cook" the meal over a longer period of time. The food has to travel through a longer stomach. This means, of course, that the pygmy mammoth is in a no-win situation. The greater it evolves in one direction, e.g. growing large to keep warm in order to solve one problem, the less able it is to survive, given the other problem, finding enough food to feed that size. Many species became extinct because of this very principle.

Flowers and Bees

I have read of the supposed problem in the contemporaneous evolution of flowers and insects. Well, I really see no problem. First, it is quite obvious that insects that lived off of pollen were around long before flowers, since pollen would be a rich source of protein to many an insect. Taking into consideration that plants of the time would not have had flowers to compete with, one can reason that those plants naturally had a monopoly on pollen-eating insects. These insects would obviously help pollinate the plants, as they do today, by carrying pollen on their bodies. It is a fact that the new shoots at the end of a twig where the bud would be tend to be slightly lighter in colour than the rest of the plant. Therefore, the insect would only have to aim for the lightest-coloured leaves in order to get nearest to the pollen. If, however, the plant delayed distributing chlorophyll to its leaves at the tip of the twig, which would give the leaf a whitish appearance, then those leaves would naturally be superstimuli for the insect. Then it becomes a simple case of which insects had the greatest desire to aim for the white amongst the leaves instead of wasting so much time searching on the wing. Insects that did this had the greatest chance of survival. At the same time, plants with the biggest and most colourful leaves (petals) attracted the most insects. As plants that are closely related to one another tend to have similar petals, any insect with a preference for one type of petal would fertilise the same species. It was then only a matter of time before some of the pollen evolved to become nectar. While we are on the subject of pollen or nectar eaters (bees), I'll mention the problem of why the workers work so hard for the females that remain sterile. The question is, Why would they have a choice? Choice comes from a large, conscious brain; otherwise, it's all inherited instinct. I would say that those mother bees (queens) that produced sterile daughters which desired to stay with and work in the hive would survive the best. The workers are, to the queen bee, nothing but extensions of herself that she has grown, like growing extra limbs. It is the queen bee that is the genetic unit of the hive. If the workers of a queen bee develop a new, superior characteristic for the survival of the hive, then that came from the genetics of the queen bee that bore them, even if the characteristic is dormant in the queen bee herself. Therefore, the queen bee will survive better and live longer to have more queen bee daughters and drone sons, who will inherit, genetically, the ability to produce more workers with that superior characteristic and survive better in future.

On Dinosaurs

It is generally thought that dinosaurs were cold-blooded lumbering reptiles. Dinosaurs evolved at about the same time as mammals, but they won the race for supremacy. Why? I don't know for sure. But it does seem that the dinosaurs may have relied on instinct rather than pragmatic thinking (relatively speaking, compared to the mammals). Therefore, this meant that the brain could be very small and the organism did not have to spend time learning before it could survive on its own. Although mammals do rely on instinct to a great degree, it is learning ability, when this characteristic is fully evolved, that seems to give them the edge. However, the dinosaurs achieved a dominant position before the mammals fully adapted to this ability. Of course, it was then very hard to unseat the dinosaurs. To live by thought, where the young have to be taught and protected by the parents, rather than by instinct is somewhat complicated and requires a great deal of time to evolve. But once evolved, this characteristic is valuable in a changing world. For the individual, thinking takes more time, whereas instinct is instantaneous. You could say that thought is slow and indecisive compared to decisive instinct, but thought is more flexible and accurate in the long term. It appears now to many palaeontologists that dinosaurs were warm-blooded. I remember reading many years ago, when I was a young man, that there was a group of dinosaurs referred to as "bird-hipped" dinosaurs. It did pass through my mind (fleetingly) that birds evolved from a small dinosaur of this type. It's like this: Bats did not evolve to have warm blood and a four-chambered heart so they could fly. Evolution does not work like that. Evolution works on the basis of pre-adaptation. That is to say that an organism has pre-adapted characteristics which it developed in a previous environment and which can be used, perhaps imperfectly, in a new environment or niche. Bats evolved from warm-blooded tree shrews, which evolved skin flaps similar to those of the flying squirrel – which, of course, only glides – but the bats took this adaptation even further and achieved true flight.

I would suggest that dinosaurs had warm blood and a four-chambered heart and that birds evolved from some insect-eating half-pint that may well have, like bats, taken to living in the trees. Why the feathers? Well, they could have been created from scales to serve as a threat display. Or, possibly, feathers evolved to provide insulation from the cold, especially at night. Since the first dinosaurs were small creatures, they were obviously more susceptible to the cold. It has been suggested that all small or young dinosaurs had feathers to keep warm, especially given the cold nights. It is very possible that the first warm-blooded dinosaurs had scales which became spines for protection. These would have slowed the airflow over the body. It was then a matter of time before the spines grew thinner to trap warm air against the skin. But instead of the spines simply growing thinner, they could become roughened along the edges, which would have slowed the airflow even more. These thinner spines could have eventually become feathered. All outer coverings – skin, scales, shells, hair, and feathers – can be used as a display mechanism if they are grown large (or long) and elaborately. Therefore, a warm blooded creature with feathers could use their feathers as a display, to threaten or sexually attract another. Feathers could also be used as wings with the ability to glide from tree to tree in search of insect prey, could have been the beginning of bird evolution.

It has been found that birds have three fingers, just like dinosaurs, but birds have a different set of three fingers. Therefore, I would suggest that the first dinosaurs had four fingers and that birds lost a finger in the process of evolving. This should help us understand what the dinosaurs were really like, as far as their internal metabolism goes. It has been found that the dinosaurs swallowed huge stones to grind up and help digest their food. They had a crop, as do birds. Also, many dinosaurs lost their teeth and developed a beak. Birds are not the only creatures who have this characteristic. It is also said that only young dinosaurs had warm blood. Young dinosaurs generated heat internally until they grew large and were able to stabilise their body temperature by absorbing heat during the day to make up for the heat loss at night. They could have had some form of temperature control, such as panting or possessing a large area of skin full of blood vessels that could give off heat during the day, if necessary, and close up during the night, if necessary. Many mammals have systems like this – again, when necessary. With larger herbivores, the fermentation of food in the stomach could produce a fair amount of heat without needing the brown fat which mammals have for heat production. This would be more efficient when so little energy existed in ferns of that time. A larger animal can consume and digest vegetable matter a lot more efficiently than a smaller one, since the vegetable matter takes a long time to digest by fermentation and, therefore, spends a long time travelling through the stomach.

One thing that puzzles me about the brontosaurus is how it protected its young. With only a tail to use as a whip, it might have had trouble defending itself. I would suggest that in order to make up for the high mortality rate of its young, the brontosaurus would have been capable of laying a great many eggs. When considering the size of a brontosaurus and the small size of its eggs, it stands to reason that a female brontosaurus would a perpetual egg-laying machine. Also, considering the huge difference in size between a newborn and adult brontosaurus, it would be surprising if the adult were aware of its young and could do anything to help them. Imagine a horse trying to look after mouse-sized young; she would be lucky not to tread on them. A further point is that although dinosaurs initially won the race for supremacy, they served as good testers for the mammals, which probably had to contend with, in addition to dinosaurs, birds, reptiles, and large spiders. It is no wonder that by the time the dinosaurs kicked the bucket, mammals were fully evolved.

What killed off the dinosaurs? I support the idea of cataclysm (as I would prefer the dramatic). It is believed that, sixty-five million years ago, a meteor or comet hit the earth and caused a disturbance, to say the least. The impact would have caused huge dust clouds, the likes of which would have dwarfed Mount St Helens' output in the early eighties. These dust clouds would have shielded the earth from the sun's rays and allowed a global winter to engulf the earth. The impact also caused, it is believed, most of the earth's vegetation to temporarily die off, which brought disastrous results to the land herbivores and, in turn, the large carnivores that preyed on them. The meteor or comet, therefore, killed off every large land creature that was living on earth. I would suggest that the survivors were either those that could hibernate for a few months or those that were hibernating at the time in the far northern hemisphere (the meteor or comet could have hit the earth in the winter stage of the year). Some survivors would be reptilian in type, able to survive many months without food. There would, of course, have been a lot of dead meat around, and small mammals, especially, could have eaten it. The large dinosaur bones would have also contained a great deal of marrow, which only small creatures could have gotten at and made full use of. There were also grubs and other invertebrates hidden in the ground. Insects would have been the first to breed and multiply as the earth began to warm up. Insects seem to survive any disaster, so insectivorous mammals and birds could have survived by eating insects. There would also be scattered amounts of greenery from place to place, but certainly not enough to feed a large animal, since it would have used up too much in energy getting from one place to another. But a small herbivore such as a primitive rodent might have been able survive if it were highly territorial and in a luckily sheltered spot. Also, there would have been seeds, nuts, and bulbs in the ground, which the small herbivorous mammals would have gone for.

Warm Blood

How warm blood evolved in both mammals and dinosaurs is a question that many zoologists would like to answer. Could there be a common origin for warm blood, such as a warm-blooded reptile from which both dinosaurs and mammals evolved? I honestly don't know, nor, for that matter, does any zoologist or palaeontologist at the moment. The strange thing is that birds' four-chambered heart is constructed slightly differently from that of mammals. The Vessels to and from the heart pass around one another differently. Therefore, although I would suggest that dinosaurs and mammals do have a common ancestor but the two types of creatures were separated at an early stage. It might be suggested that they both evolved from a reptile which inhabited cool regions of the world. This creature may well have used hibernation to begin with, but it would certainly have been a meat eater that relied upon consuming enough protein to generate enough energy to keep itself warm and engage in the high activity of hunting. But what would it have fed on? The only protein that was available in cool climates, particularly in winter, before any warm-blooded creatures evolved were fish from freshwater rivers and lakes or, perhaps, the sea. Perhaps they ate invertebrate shellfish on the shore. Could they have been water-living hunters? They would certainly have needed to generate heat to swim, since water takes heat from the body very quickly. In the Galapagos Islands there exist some seagoing iguanas, which live off of seaweed. They dive into cold water to feed and then heat themselves up by basking in the sun. After storing up energy this way, they dive again. Since these reptiles live off of seaweed, they do not have enough energy to heat the body internally, even if evolved the mechanism to do so. Perhaps the answer is as follows. A small reptile, which, as you know, can lose heat very quickly, could dive for fish (which have a high amount of protein) in an in-and-out fashion, thereby not losing too much heat in one go. However, if some heat-generating ability were to appear in a particular individual, then that reptile would be able to spend more time in the water, gathering more food to generate more heat and engage in a greater amount of activity. There is also the night-hunter theory. That of a reptile that hunted at night in certain areas such as a sandy desert which can get very cold. It could start off by hunting in early evening while the earth is still warm, and then retreat as soon as it got too cold. If any reptile were to evolve an ability to generate heat, it would be able to hunt and remain sexually active for a longer period throughout the night. Of course, this would only be an explanation for the kick-off for warm-blooded evolution. Once the blood became warm and the individual became highly active, it would certainly diversify elsewhere.

Chapter 2

Mammals and Primates

A Few Notes on Mammal Evolution

On Hair

There are questions about how hair evolved in mammals. Given that the reason for hair's evolution is to provide insulation, how could an animal be insulated before hair was fully evolved and developed? I would say that the most likely explanation is that the scales of the pre-mammal grew as spines for self-defence. The longer they grew (growing into quills, perhaps), the greater the defence. Of course, the thinner and more compacted the spines became, the slower the airflow over the skin and, therefore, the more heat retained. Therefore, any pre-mammal that needed to reduce its heat loss would, in evolutionary terms, grow thinner strands of more compact hair. By that time, the creature would have become more active and fled to escape its enemies rather than rely on defensive quills. It is said that porcupine quills are adapted hair, I believe that hair adapted to become quills, although this could work in reverse from time to time. As a side note, the most primitive insectivores, tenrecs (shrew-like creatures), which live on Madagascar, are known to have spines in their fur. Also, the echidna, an egg-laying mammal (a monotreme), has spines.

On Mammary Glands

Another problem for evolutionists is determining how mammary glands evolved. I give the following as a possible explanation. It is known that some birds in dry places bring water to their nestlings by soaking their breast feathers in water before flying back to the nest. I would speculate that the early pre-mammals did the same thing, but that they carried the water in their underbelly's fur. Therefore, the young evolved the strong habit of licking the mother's underbelly. Of course, there would be a selective advantage in efficiency if the mother drank plenty of water and then sweated profusely when back at the nest site. I believe that this advantage evolved over a period of time. With the minerals, fats, proteins, and vitamins that sweat contains, the mother would not need to carry so much food back to the young to keep them healthy. It was only a matter of time for milk to evolve. However, the milk may have evolved from the oil glands, not the sweat glands. This is something that biologists should investigate.

The echidna and platypus (which are both monotremes) do not have teats on their bellies. They eject milk through the pores of their skin, and then the young lick the milk from the skin. This method can be somewhat wasteful and inefficient. Still, if some of the pores or glands had become specialised in excreting milk, then the young could localise their licking. Since mammal mothers are highly active and moving about the nest, it would be a selective advantage if the enlarged group of pores or glands protruded and, thereby, became easier for the young to hold onto. Also, there would be less spillage. I believe that the mammal teat evolved this way.

There are two types of monotremes (egg-laying mammals): the platypus and the echidna. The platypus rears her young in a nest. The echidna carries a single egg in her semi-pouch (consisting of two folds of skin). Of course, the common ancestor of monotremes and modern mammals would have had a full set of teeth as well as a more conventional insectivorous lifestyle. The monotremes of today do not have those teeth because their lifestyles became very specialised over many millions of years. I wonder whether the two types of surviving monotreme mammals give a crude representation of the very early evolution of today's two types of mammals: the placental and the marsupial. Could it be that the platypus's nest building marks the start of the placental mammals and that the egg-carrying echidna marks the start of the marsupials? It would be interesting to know if a biochemist who can perform DNA dating can find out how long ago these two species separated. It could very well be that these two monotremes are separated by a greater or similar distance of time as the marsupials and the placental mammals.

Primates

The first question I would ask when contemplating primates is why they have forward vision. The general explanation is that they live in the trees and have depth perception, or stereoscopic vision, for jumping from branch to branch. While this would be very useful for a tree-living creature, squirrels managed for quite a few million years without forward vision, let alone stereoscopic vision, and no inclination to evolve in this direction. I would suggest, after looking at the range of creatures throughout the world, that there is no exception to the rule (except the primates?), that all hunters, or meat eaters, who rely upon eyesight for hunting have forward vision. However, all non-hunters have side vision or all-round vision unless their herbivorous activity is a recent evolutionary trend, such as with the giant panda.

Once forward vision has evolved, especially binocular vision, it may be too good for a species to give up. It would appear that an herbivorous tree-living creature would be better off if it could see a predator (especially a bird of prey) coming from a distance and have a good start at getting away than if it had forward vision and was caught unawares. It is quite obvious that once a creature in the trees sees a predator coming, it is easy for it to hide or take evasive action, especially if it is smaller and can hide in the denser parts of a tree, where a bird of prey is not able to fly, or move along smaller branches that cannot hold a heavy tree-living carnivore. A hunter in the trees not only has to outmanoeuvre prey but also requires quick decision-making skills to determine whether or not a branch is safe (i.e. it may be dead or weak) or how strong a particular species of tree is. A creature of this type really has to think fast.

Colour vision is a classic primate characteristic. It requires not only colour cones in the eyeball but also more brainpower to analyse the information coming from the large number of cells in the retina. Why did colour vision evolve? I must confess that fruit eating had a great deal to do with it. Fruit, when unripe, is normally green, the same colour as the surrounding leaves. Upon ripening, it changes colour and (in a manner of speaking) makes itself conspicuous. However, colour vision would not go amiss for a creature when hunting insects. Anyone who has watched wildlife programmes on TV will note that camouflaging insects are far easier to make out when against something multicoloured than when on a piece of monochrome fruit. (Sometimes this characteristic may be contradicted). Fruit wants to be eaten in order to spread its seeds, or progeny, out into the wider world. Fruit, like meat, is very digestible; therefore, any carnivore would not turn down fruit if it were easily available. However, fruit eating, I would suggest, was not a full-time occupation for our ancestors any more than it is for the modern gibbon.

One significant thing about primates is their social dependence. Having forward vision instead of all-round vision means that a primate needs many pairs of eyes on watch. (Of course, if an animal is big enough and has no predators, then it may do its own thing.) Social dependence, however, presents many problems. One of these is competition amongst those who are interdependent. Forward vision is a good thing, but it has its price. I would now suggest the following as a summary.

Early shrews that took to hunting in the trees relied more on vision for seeking their prey. As it is easy for prey creatures to jump from branch to branch, they do not leave continuous scent trails as they do on the ground. Therefore, a reduction in the nasal cavity (muzzle) would give a better view below without compromising the creature's ability to hunt. Early tree shrews first prey would be tree-living insects. I would suggest that as primates grew in size, they ate birds' eggs and fledglings. Then, as the primates evolved to the lemur stage, they ate full-grown birds. Of course, in every stage of primate evolution, these animals, especially the relatively lazy and non-aggressive amongst them, retained the option of being herbivores. Fruits and nuts would be the first option for the non-hunter, since digesting these foods does not require a specialised stomach. Fruit, in a manner of speaking, wishes to be eaten so that its seeds will go through the animal and be deposited on the ground along with a nice lump of fertiliser. Nuts are a rodent's option. When rodents bury nuts in the ground to store them, they are unwittingly planting the next generation of trees if they forget about some of those nuts and leave them buried. Primates can always take nuts as a secondary option, even though they are somewhat indigestible compared to fruit. Please note that in cool climates where mammals hibernate, trees are of the nut-bearing variety, since rodents, e.g. squirrels, coexist with them and store their nuts. Very few naturally occurring trees in the northern zones actually produce fruit. Since primates do not hibernate and therefore only inhabit warmer climates, they do not store nuts. They would eat the nuts as they gathered them. Therefore, the trees in warmer climates would be better off producing fruit. As primates tried and tested more trees, some omnivorous primates gradually evolved to become full herbivores. The advantages of being an herbivore are that it is less dangerous and can support more individuals who inhabit a given area of land. Also, herbivores have only to watch out for predators and are in less danger of being a hunter's prey. I believe that virtually every stage of primate evolution occurred because of the primates' predominant and continuous hunting, or meat-eating, option.

Chapter 3

Apes

I believe that human evolution began when the ape achieved an upright stance. The first apes, I believe, evolved from an advanced lemur-cum-monkey with the habit of jumping from tree to tree in an upright position, grabbing hold of the tree trunk instead of the branches. This would be a safer way than running along the top of the branch and then jumping to another branch top, especially given that when a primate gets too large, its ability to balance can become impaired. This adaptation would encourage an upright position. The early non-brachiating apes must have evolved as far as this evolutionary stage. In Madagascar, a species of lemur called a sifaka has this habit. When on the ground, it jumps along on two legs instead of running or walking on all fours. This particular lemur has proportionally long legs which it uses to jump from one tree trunk, flying head first through the air and then rotating its body with its legs forward, and then break its fall against another tree trunk. The sifaka, having evolved with very long legs and an upright stance, finds it awkward to walk on all fours. However, the sifaka is so specialised in tree jumping that it can only manage to jump or hop on the ground. I would say that the first apes (which were non-brachiating) were also bipeds when on the ground. I am not suggesting that they jumped along like a specialised sifaka; rather, they walked on two legs. These apes, although relying on their feet to break a fall, would still feel more secure if they aimed for the part of a tree just under and beside a branch. This way, they could use their hands to grab hold of the tree. An ape performing this action would tend to land on a trunk with its hands above its head, which would be somewhat awkward for a non-brachiater. But with evolution being the way it is, it would favour the well-endowed creature whose arm could rotate above its head.

I would suggest that as the non-brachiating apes expanded to many species, a lighter, more active ape that was adapted to grabbing tree branches with its hands alone (rather than using its feet, as well) developed. I believe that some of these apes developed arm-over-arm mobility (brachiation) in the trees as they become more active and faster-thinking. A quick-witted gibbon can travel faster in the trees by brachiating from branch to branch rather than from trunk to trunk. It was about ten million years ago when the ancestors of the gibbons and the rest of us apes went our separate ways. According to the fossil record, brachiating apes came into existence about fifteen million years ago (which means, obviously, that brachiating has been around for at least that long). I would also say that all the surviving apes today are, in fact, bipeds who have degenerated into knuckle-walkers, except, of course, for the gibbons and siamangs, who are specialists in the ways of apes. All great apes practise knuckle walking, not the palm walking which monkeys use. I believe that when the evolving brachiating apes came down from the trees, they, from time to time, mixed knuckle walking with bipedalism. I honestly believe that bipedalism is the natural way of all brachiating apes but that the great apes have reverted back to the quadruped way. I shall explain my reasons for saying this later.

There are many theories as to why brachiation survived so well. The most common explanation is that primates used it to get fruit from the branch tips. However, monkeys are reluctant to brachiate. Monkeys have shown no movement towards brachiation and no tendency to evolve brachiation (apart from some spider monkeys in South America, which have a very lightweight build) even though fruit is an important part of their diet. In any event, I believe that brachiation, once evolved, is an adaptation that allows high-speed mobility (which gibbons have perfected). At the same time, brachiation enables an ape to carry food in its feet to its home base. Gibbons are known to do this. Since food sharing is not normally the habit of an herbivore, I would suggest that meat eating was the original start to this habit. Then, the special condition of monogamy in a primate whose mate was lumbered with helpless young kept the habit going, even though fruit may have become the main source of food in such a situation. The ape, in these circumstances, would still defend a specific territory. The female, on her own and encumbered with a baby, would not be able to defend a territory by herself, so having a mate to do this for her was beneficial. Defending a territory means that one doesn't have to compete with one's own kind for food. Plus, the territory is not depleted of its food supply by way of too large a population. *Humankind in this day and age could certainly learn from this.* If an ape was a meat eater, then there was a high probability that it carried meat home to its mate and young between its feet. It could well be that the ancestors of all brachiators were meat eaters, but only the gibbons retained, at least partially, this habit. It is highly probability that, because gibbons lived in the trees, birds were their main prey. Gibbons are known to catch birds in mid-air. I do not know if they placed their prey between their feet before catching hold of another branch (the textbooks I have read did not provide this information, as it is not yet confirmed), but this would be the ideal thing for a meat-eating brachiator to do. I have never heard of any monkey doing this, and I don't know whether monkeys are capable of it. I would suggest that brachiation and the meat-eating way marked the start of human evolution, as I shall explain in due course. First thing in the morning, most people feel like eating an egg and soft cereal. I would say that for millions of years, this was the easy option for creatures who were feeling a bit sluggish within a defended territory. There does seem to be an obsession for studying the chimpanzee and gorilla in order to understand humankind's ancestry. I would say that the gibbon's lifestyle, (more so than the modern African apes' lifestyle), is more like the lifestyle of our own ancestors, since, as I shall explain later, the African apes have moved away from the true ape way of life. I believe that we would understand more about our ancestors if scientists received more funding to study gibbons. (before they become extinct that is).

Gibbons and the Evolution of Early Humankind: Black Bad; White Good

There is a species of gibbon in East Asia consist of male adults who are black, females who are fawn (blonde), and young who are white. The male adults are very aggressive towards any adult male intruder such as a black ape (which is monogamous and mates for life). Gibbons are tolerant of the young, that is until a young male turns black and, at last, perfects the territorial ritual call. At this point, the adult male gibbon ousts the newly matured young male from the territory. It is generally accepted that men prefer blondes and that women prefer men who are dark. Also, we associate whiteness with innocence. I wonder if this is significant to the gibbons' case. It would appear that the adult gibbon has a thing against blacks, even though he himself is black. (This is not exactly liberal-minded or politically correct.) It has been reported that some male gibbons are fawn in colour and that some females are black, but this is somewhat rare. It may be due to an aberration like homosexuality. Also, many species of gibbons do not all conform to type. In some types, both genders are black and only the young are white. In other species of gibbon, black and blond appear in both genders. Psychologists often wonder why people regard blond hair and blue eyes as almost magical. I say that these traits have a calming and soothing effect on the human psyche. Darkness has an opposite effect, perhaps. Maybe we evolved from a species of ape whose colour was more fixed to a particular sex. I shall delve into this later.

Stan Gooch pointed out in his book *The Guardians of the Ancient Wisdom* that gibbons have a twenty-eight-day menstrual cycle, a period far closer to the human female's than to any other ape's. This may mean that our own sexuality has, for the last ten million years, been influenced or governed by the moon. It cannot be just a coincidence that it takes twenty-eight days for the moon to orbit Earth. Not only has the human female been influenced by the moon, but also those men who were the most sexually active at the right time (i.e. relative to the period of the moon's orbit) would have a selective advantage. The human gestation period, 280 days, is exactly 10 lunar months. Is this a coincidence? I wish to make it clear that I do not say that humankind evolved from a gibbon or even from an ape that looked like a gibbon. What I am saying is that the apes from which humankind evolved had a lifestyle and habitat similar to that of the gibbon. The ape that we evolved from, although a brachiator, was not as specialised a brachiator as the gibbon is today. Remember that we have been separated from the gibbon for about ten million years and that the gibbon has, no doubt, been evolving to suit his niche for a very long time.

Neoteny

Neoteny is a word often used in human evolution. It indicates a slowing up in the development of adult characteristics, also growth does continue. In other words, neoteny describes an immature or undeveloped organism that is fully grown. The human being is considered to be such an animal – in other words, a person is an immature ape. However, in such cases, the individual is still able to reproduce, thanks to overriding characteristics of the neotenous effect. Neoteny is known to be a factor in many cases of evolutionary adaptation. The overriding factor may occur in more than one characteristic of a particular organism. It is neoteny that increased the size of the human brain, although there is more to intelligence than having a big brain. (A whale's brain exceeds a human's in absolute size, whereas a squirrel monkey's brain exceeds a human being's in relative size.) In human evolution, neoteny is responsible for the shape of the foot, the desire to play, the upright large head, the hymen, fair colour, a reduction in the size of teeth and jaws, etc. It is also possible for neoteny to work in reverse. There exists a species of amphibian in South America called the axolotl. These amphibians reach sexual maturity while in a tadpole state and living in a lake at high altitude. They possess gills and therefore breathe underwater. However, if the oxygen level in the lake diminishes, the amphibian develops lungs, which is a reversal of the neotenous effect. Just because the genes have been turned off does not mean that they have disappeared altogether. The genes could be reactivated at a future date.

Human Apes Becoming White

Given that brachiating apes evolved in India, and also given that I believe that Neolithic man was a cold-weather animal, I would suggest that Western Asia is the place where the central line of human evolution appeared. It is far easier to live in Africa if a species can breed fast enough to offset the loss of its members to the many diseases that afflicted the young and old. In contrast, in a cold climate, there is very little food, especially in the winter months, so Neolithic man would either have had to hunt in the extreme cold or store away food for many months.

I would suggest that between seven million and five million years ago, some apes found that the world became cooler, the sky more cloudy, and the trees sparser and more isolated. A species of ape living in that time period might have already adapted itself to ground living in the few open spaces. It's this species of ape that would have expanded once a profitable change in climate occurred. There would, I suggest, be an advantage if the young remained white well into adulthood, since this would cut down on interspecies antagonism. The young would gang together with their father and siblings to go on hunting expeditions – on the ground, perhaps. Probably a mere delay in black maturity would occur. This would give the young, in a fully grown state, the time to help the parents bring up younger siblings, gain practice in and knowledge of hunting, use territorial defence and child-rearing before finding their own territory, and rear their own young with the ready-to-use skills intact. (This is known to occur in many species of birds.) Supergroups, or multiple families closely related to one another, of apes would gang together for the hunt. This, in turn, could indicate a double meaning for the concept of territory.

There would be a group-hunting territory for the supergroup and personal territory for the family unit. This could occur even if the adult males were black, since it would only mean hunting to gather and returning to a personal nest in the trees afterwards. Therefore, several black adults would not have to tolerate one another for any significant length of time. This may sound a little familiar to today's reader. However, by the time these apes lived fully on the ground, away from the trees, and had only each other for protection, they may have needed to be all white (or blond) when fully mature in order to stay together all the time. (But adult whiteness may have gradually evolved first while they hunted on the ground and nested in the trees.) This would mean that their territorial instincts and aggression while hunting remained intact to protect themselves against intruders (even if the intruders were white. The fact that an intruder is a stranger may be enough to arouse aggression), but not against their own kind. Just as the parents had great affection for, protective feelings towards, and bonds with their white young, they would have had these feelings for the group, *so long as each member was familiar and white in colour*. One word of warning: Just because the need for antagonism towards black apes disappeared does not mean that the response disappeared from the ape or hominid gene pool. The longer a characteristic has been in existence in the gene pool, the harder it is to get rid of. Black male adults, fawn female adults, and white young may have existed in our ancestral evolution for many millions of years.

The alternate view is that *Hominidae* were tolerant of blacks so long as they were familiar, but not if they were strangers. The reader might be aware that black people love their friends and family but do not necessarily like black strangers. This would explain why black people do not have stable societies, although they fare well in a white society. A similar thing may be seen in white people, since many a white person has black friends but still does not want to live in a black society. In any event, when human beings turned blond, this enabled them to live amongst blond strangers, which therefore allowed larger groups to develop.

A biped ape named *Orrorin tugenensis,* which is believed to be six million years old, has turned up. At one time, I thought that the separation of human and chimp was four and a half million years, but now the DNA evidence has changed, showing that human and chimp were separated six million years ago (the goalpost has been changed). In other words, evolutionary scientists have changed the evidence to suit the theory that human beings became bipedal only after separating from the apes.

I believe that the apes which gave rise to humankind lived on the Indian subcontinent, although no apes have lived in this region for some time. I would suggest that our ape ancestors might have had something to do with this. Let's suppose that our ape ancestors had hunted other ape species in the trees and lived off of ape meat. At a time when their prey became scarce, they supplemented their diet by hunting other prey on the ground, which would not be too difficult. This would mean that if the early apemen preferred ape meat over any other, they would have been able to survive even if their ape prey became scarce, since apes are very slow breeders. Therefore, the apemen's numbers would not be reduced when ape prey became scarce. This would mean that when the ape prey was reduced to near extinction, the apemen would still put pressure on this prey and drive it to total extinction. This would explain why the apes were wiped out on the Indian subcontinent. It also explains the apeman's pre-adaptation to hunting on the ground, which would be useful during such a time when the environment changed and the trees began to disappear.

Ground Living Started

Say that some of these biped apes were specialised carnivores and, given that one habit can lead to another, started hunting on the ground. I would like to point out again that the gibbon, when walking on the ground, practises bipedalism (walking on two legs). Although this action is somewhat ungainly, it must be remembered as the habit that set evolution on its way. The gibbon is a far more specialised brachiator (which means that it sacrificed the efficiency of other means of motion) than the common ancestor of other apes and human beings, who, I suggest, were more graceful and capable on the ground. Also, the hands of the gibbon are specialised. They are almost like long hooks, rather unlike the more dexterous hands that our common ancestors possessed.

When the first apes came out of the forest, they may well have been well-adapted bipeds able to walk almost as well as the modern human being, given that for many millions of years, they periodically, inside the protection of the forest, came down from the trees to explore the ground. They may well have lived on the ground for some time in isolated spots and developed adaptations before the environment changed. There is in the theory of evolution the factor of pre-adaptation. This is when an organism adapts to one environment but can use the adaptation in another. Also, a small group of animals can adapt to an unusual, isolated environment. If those environmental conditions become widespread, then the local group's population expands outwards to fill it.

The first hominids to come down from the trees may well have adapted, as many birds have done, in such a way that the young males, even though they turned black, may be tolerated by the alpha male because the males where familiar, helped their parents hunt and look after the other offspring. (Please note this would not only be psychological change in the alpha males but also a change in the young males as well. Which is significant for future reference.)

Ground-living bipedalism, as one would assume, is somewhat dangerous. From time to time, as is the case for present-day apes, the early hominids were forced back up into trees when carnivores crept up on them on the ground. Under such circumstances, it is not unusual for apes to create a threat display. The general behaviour of gibbons and many other apes is to show a threat display by shaking the tree branches. This would be done by standing on a bough and shaking a nearby branch.

Let's suppose that some poor chap, from time to time, was observed and subsequently chased by a carnivore and was unable to make it to the nearest tree. His instinctive reaction, I believe, would be to pick up the nearest object, say, a fallen branch, and shake it in front of the predator. I am not suggesting that a large predator would be instantly afraid of a three- to four-foot ape with a stick in its hand. However, if you put yourself in the position of a predator, able to raise only one paw into the air at a time to defend yourself from a thorny branch, you see that it could be somewhat off-putting. You instinctively know that you could lose an eye, have your tender nose slit by very sharp thorns, or be heavily bruised if you were struck by a heavy stick.

A wounded predator is not much good to himself or his family. As the saying goes, "He who hunts and runs away is alive and healthier to hunt another day." On such an occasion when the carnivore backed down, the apeman would gain an extra bit of confidence, knowing that the seemingly all-powerful carnivore was not quite so powerful, after all, and is capable of being beaten. This would not only give the chap a boost in the social order but also give other apemen confidence. If one of their own kind see off the carnivore, then, it would stand to reason, the group stood an even greater chance of chasing predators away. Having a higher social standing in the group gives a male a greater chance of breeding; i.e. he has first choice of the available females. I would also point out that although carnivores will kill anything they can get at, they usually specialise. And since apes are slow breeders, any carnivore that specialised in killing apes would either hunt an ape species to extinction or starve if the apes adapted and defended themselves against general slaughter, unlike the zebra and wildebeest. I do not say that every ape that waves a stick at a predator will win the day outright, since there will always be the "idiot" who tries the threat display on a larger than normal carnivore, but I do say that doing so would give many a apeman a better chance of survival and a bit more confidence to hunt farther away from the trees. Chimpanzees are known to hunt out in the open when they have to, although I would point out that, compared to gibbons and also our common ancestor, the chimpanzee has lost his "bottle" and is not so ready to put up a sustained mass defensive attack.[1]

However, it is common for primates that live a more dangerous life in the African bush, namely baboons, to "honour" an individual who risks his own skin to warn the group of danger. One such show of honour is when a high-ranking member of the group grooms him. It stands to reason that those biped apes who were out in the open and carried a threat display (weapon) around with them would stand a better chance of survival than those that did not. To put it another way, those biped apes that felt naked without a threat display (weapon) in their hands would survive the best. I suppose it is obvious why the cowboy feels "naked" without a weapon (gun). This is not a coincidence.

If you are socially dependent on your tribe, then it is worth taking a risk and "having a go" at a predator. When you do so, you face and attack the predator, which makes things risky for both predator as well as the prey. Even if you get wounded or killed, the predator realises that attacking your tribe is more dangerous than attacking another tribe or another species that prefers to run. If, in getting yourself wounded or killed, you also manage to wound or kill the predator, then this would give the predators with an inclination to attack your tribe a lower chance of survival than the predators that hunted elsewhere. However, if you run away, you have your back to the predator and are therefore seen as a relatively harmless target. Because of this, the predator will know that your tribe is easy pickings. Also, if you make a counter-attack and get killed, then your family could flee to safety and therefore survive, even though you didn't. However, if your family's numbers were reduced after making themselves easy targets on the run, then the survivors who are socially dependent would not survive anyway. It's then a matter of

---

1 It is known that when a group of chimpanzees are threatened by a carnivore, e.g. a leopard, the lookout normally runs up the nearest tree before giving a warning. If there is any defensive reaction, it is by the young mature males. The alpha chimp generally remains safely in the tree. For this reason, I do not believe that studying the chimpanzee will tell us too much about our ancestors.

having enough individuals who are willing to have a go and present a concerted mass defence or threat display.

If a mass threat display were practised in a synchronised rhythmic beat, this would give to the predator the instinctive impression that it was up against a large singular animal. This, I believe, is how using weapons, and humankind's love of beat music, evolved. If a crowd of people wish to show their displeasure to anyone, they slowly clap their hands, which creates a feeling of strength and unity. It's amazing that a large number of people can synchronise their clapping in such a short period of time. I would suggest that this is not a coincidence. Likewise, if you wish to show pleasure, then clapping your hands out of synchronisation is the norm, perhaps because this is the opposite of a threat display.

Think of a film showing a marching army moving towards its opponents. When the soldiers move their shields or weapons in unison, the army looks like a large beast or unstoppable machine on the attack. You would be very intimidated if you were in its way. This is how an animal perceives a group of other animals that are making a threat display. Note that an animal makes no distinction between what it thinks and what it feels. It feels itself faced with a huge animal or machine, not a group of individuals. It is possible that apemen waved long sticks at predators in order to put up a mass defence, much the same way that pikes were used against the cavalry during the Cromwellian civil war. Enough men pointing long sticks at an enemy creates a form of defence similar to the hedgehog's or porcupine's. I cannot see any enemy daring to charge such a mass defence. You can now understand why such a thing would make dependency on one's pack essential. Of course, any hominid that strayed from the pack would be a relatively easy target for the carnivore. Those hominids that kept close to the pack and felt afraid when on their own would also survive better. The early years of humankind's evolution did not suit the loner or individualist.

When I saw the film *Aliens,* I thought that the aliens themselves did not appear all that dangerous (considering the firepower available to the men and women who were in the situation) compared to, say, lions, tigers, and many other large carnivores that lived now and also many millions of years ago. So, why were the aliens so obviously deadly to human beings? The real reason is not their speed or physical strength, but the fact that they had no fear of people or of creeping into their territory – in this case, the spacecraft. The fact is that modern human beings are used to thinking that animals are automatically afraid of them. Like the people in *Aliens*, we are, therefore, somewhat overconfident.

Robert Ardrey pointed out in his books that predators had to learn to be afraid of humans. Those predators that had a go at apeman the pack animal stood less chance of survival than those that left the apemen alone and/or had an inbuilt fear of him. It must have been that our apemen ancestors were formidable in their early years. Just imagine a large pride of lions descending on a lone man who had a totally fearless dedication. Even with a rifle, he would still need time to reload between shots. Just imagine how an apeman would handle the situation. He certainly wouldn't attempt it alone. (This may be a warning to people when they visit new environments yet unknown.) I saw the film *The Ghost and the Darkness,* which is the "true" story of man-eating lions that appeared during the making of a railway bridge in South Africa. These lions killed many people. The obvious point is that they had little fear of people. Even a professional hunter fell afoul of them.

Robert Ardrey made a strong point about the phenomenon of enmity and amity between groups in the wild. Enmity equals hostility; amity, friendliness. The general principle is that, in a tribe, while individuals may compete with one another, there is a need for social cohesion. The greater the threat or antagonism from outside the group, the greater the bonding within, which creates social cohesion. During World War II, the feelings that the British people had for one another were strong, as all were bound by the mutual joy of togetherness and had a seething animosity for the enemy. Even without an enemy, the British make for a reasonably stable society. But would even the British people be so sacrificing towards one another without an enemy? The tribal human animal certainly had enmity and amity for other tribes as well as for members of its own tribe. This issue is very relevant in today's world, but it does not mean that all people need enmity if they are to have amity.

I personally believe that chimpanzees, gorillas, evolved from these ape men (Australopithecus), since chimpanzees, and gorillas, use materials from their immediate environment to show a threat display. But they have lost their "bottle" or, perhaps, separated from the main line of hominids before they evolved to have many hominid characteristics. The old stories about gorillas is that they had a nasty habit of tearing people limb from limb (although this has been contradicted by modern zoologists who study the gorilla). This was probably the case until the white human being came along with the gun. At that point, any gorilla behaving too bravely was shot. This meant that gorillas that were reluctant to do battle with human beings, stopped providing a threat display, or got out of a human being's way in the first place. This made them more likely to survive. Therefore, it was the white person's gun perhaps that made the "nice" gorilla of today.

The strange thing about chimps is that they will occasionally go hunting in a pack. It has been noticed in many clans that the one making the kill is held in high esteem by the rest. The successful hunter will share the kill with the other chimps, but even the alpha chimp does not try to steal food from the hunter. Instead, he waits passively until the food is offered. Is this ritual something reminiscent of the past, or of, say, the common ancestor of apes and human beings? According to the fossil record, brachiating apes arrived in Africa fully evolved. Also, Asia (India), not Africa, is the place where most apes thrived. I therefore assume that Asia is the homeland of apes and human beings, even though I know I espouse a minority view here.

When animals migrate from one place to another, they are looking for new territory to settle in. They do not necessarily know where they are going (unless it's during annual migration, of course.). Therefore, the idea that apes, like Moses, went to Africa through Egypt in a single trip is nonsense. Animals do not go for exoduses. They only spread out in a circular fashion when they see good territory. Therefore, the apes would have spread out in all directions, north, south, east, and west, if they detected new territory.

It is known that at about seven to five million years ago, the world's climate began to deteriorate before eventually entering the Ice Age, which occurred about 650,000 years ago. By this time, hominids could have very easily wandered into southern Russia around the Caucasus, and other places as well, when the climate was moderate. However, when the climate turned cold, the higher regions between India and the Caucasus Mountains would have turned cold before the lowlands. The hominids would have found themselves cut off. Necessity is the mother of invention, as it is said. This may have forced the hominids to protect themselves from the cold by wearing animal skins. Did hominids use fire by this time? Who knows? It is during one of these periods that some humans became trapped behind the Himalayas, which had frozen over. This may have happened many times in the past. Many an ape tribe would have perished. But at one time, perhaps, a very clever group would have used fire and skins to keep themselves warm, especially at night. Using skins could have been a habit to disguise themselves when on the hunt to get close to their prey.

It is this isolation from the warmth of India that would have accelerated the evolution of humankind, since having a high-powered brain would have been a great advantage in light of very little competition from less intelligent tribes, which would have died out. Cooking would have taken off when humans began defrosting stored food. They may have overheated the food beyond the point of its defrosting, which would have cooked the food. This would have made the food very nutritious. Humankind would have known fire by having witnessed forest fires. Being very opportunistic, human beings would have worked out how fire spreads from one combustible material to another. Early human beings probably would have found that there were good pickings after a forest fire and perhaps experimented with burning embers. One can never really know how humans actually discovered fire and learnt how to use it. One can only speculate.

Chapter 4

Humankind Kicks Off

Water, or Aquatic, Ape

There has been much said about the aquatic ape, as human beings exhibit many characteristics which could only be explained if our ancestors lived an aquatic life. For example, humans have streamlined hair growth, webbing between the fingers, and a love of water and swimming. (I know it can be said that people enjoy flying. Someone might ask, "Does this mean that humankind used to fly?" Well, yes, sort of. When an ape leapt from tree to tree, it would have felt weightless for a brief moment, feeling the sensation of flying. I believe that a creature does not always take to a way of life out of necessity but because it feels good about it.)

One other puzzling question is that Caucasian humans have lightweight bones, which are ideal for swimming, while other races such as South Saharan African and Far Eastern Man have heavier bones, which gives them a disadvantage in the ability to float.[2] Could it be that Neolithic

---

2 This I have read from time to time in books on anthropology. I have heard some objection to this point, but those people who object are not likely to produce any proof, even though they may be the only ones with the power to find it. They would probably say that I am wrong if only one so-called black person were found to have lightweight bones. Counting the exception in order to prove the general rule wrong is of course bigotry.

man was more aquatic than Neanderthal man? More about that later. I would say that the greatest proof that humankind went through an aquatic stage of evolution is that very young babies put in water swim naturally. That is to say that they will automatically hold their breath and feel at ease under water, without any sign of fear or panic. (I wonder if this is only true of Caucasian babies, since I have only seen Caucasian babies to do this. Do South Saharan African or far eastern babies react the same? If not than the water adaptation may be exclusive to the Caucasian line.) If you did this with any ape, then the ape, if not immediately rescued, would drown.

I suggest that at some time in the past human beings hunted large mammals in the water. Let us suppose that some primitive hominids waited in the river for some migrating antelopes to cross. They could swim under water, come up by the swimming antelope, grab hold of it around the neck, and hold its nose under water until it drowned (or they slit its throat with a stone knife). Of course, these hominids would have had to make sure that they were not hunting something bigger than they could hold onto, i.e. biting off more than they could chew. Many animals, although good swimmers, are helpless in water and are therefore very vulnerable in it. Also, I find it most interesting that people may eat sea (saltwater) invertebrates, e.g. shellfish (crustaceans, et al.), but are repulsed by land invertebrates. I therefore wonder whether we are adapted to seawater or that sea (saltwater) invertebrates, unlike land invertebrates, do not carry the bacteria that harm us. I suggest that the water ape appeared after four and a half million years ago, since the other apes show no inclination towards water whatsoever.

Apeman the Hunter

It has recently become fashionable to ridicule "man the hunter," especially amongst women's-lib types (or any "new man" wimp who has jumped on the bandwagon of women's lib, since women's lib is the "flavour of the decade"). These people say that humankind, or "person-kind," could not have been hunters. "How can a four-foot-zero creature hope to catch up with a fast-moving herbivore?" Also, they say, the size and strength of early hominids makes the idea quite ridiculous. Apemen, therefore, must have been a glorified scavengers. If you observe animals in the wild, then you know that unless you are a flying vulture, you are more likely to be beaten to the kill by other scavengers that are a lot bigger and a lot faster than you are. It has even been suggested on TV that apemen in Africa lived off the marrow in scavenged bones. I doubt whether this would have encouraged the development high intellect. It could have been that when apemen made a kill, they broke open the bones to eat the marrow in order to get the most out of the meal. Other animals, except for hyenas, wouldn't bother with marrow. However, Africa is not the place that seemed to encourage intellectual development. It has been noted by geneticists that a hot climate produces a high variety of evolving germs, both bacteria and viruses. This means that any organism must have the greatest variety of antibodies and bodily defences possible. This can be achieved by prolific breeding as well as a great deal of cross-breeding with other individuals or groups, with an emphasis on survival of the fittest vis-à-vis resistance to germs and with less emphasis on survival of the best, i.e. the most intelligent. Of course, immigrants from Asia would have brought their Asian-evolved intelligence to Africa to add to the gene pool of those who were already there. It is one thing to evolve a higher intelligence and that of selecting out genes that have already evolved. It does appear that hominid skeletons have mostly been found in Africa. Very little investigation has been done elsewhere.

If the main question is, how does a creature of four-foot-nothing hunt down a large herbivore?, then the answer lies not only in the aquatic explanation but also in the bowler theory. At many hominid sites, carefully chipped balls of stone have been found, mostly in pairs. It has been suggested that hominids could have tied the two stones with leather thongs, thereby creating bowlers which, if thrown at the legs of even a large herbivore, could bring down a large creature. One might ask, Why go to all the trouble of making very round stones to use as bowlers since a couple of odd-shaped stones would do? Well, no, they would not. It is very useful to bounce bowlers off the ground in a more or less predictable manner if one's throw is somewhat short. I do not believe that hominids would have made these round stones just to throw them once at an animal, since one stone alone can easily be lost. However, two stones tied together with a long thong makes sense, especially if the thong was dyed with a colour. Also, hunting is not a simple case of running after prey. Many hunters rely on stalking and entrapment. The early hominids may have used fire or loud noises to frighten their prey and lead it down a prepared corridor of trees and makeshift fences into a catchment area. Also, I remember reading that humans, when physically fit, have stamina greater than many prey animals. When hunting dogs hunt down their prey, they tend to chase it in a circle, keeping the prey always to one side. This forces the prey to run in a circle. The few dogs in the chase have to make an even bigger circle to contain the prey. The rest of the pack, far behind the others, take the inside circle and do not run as far as the prey. Pack members in the rear relieve hunters in the chase. In the long run, once the kill is made, each member of the pack runs a shorter distance than the prey. Could early hominids have used this approach to hunting? Doing so would have taken some planning. Could that have been the start of primitive speech?

Women's libbers say they cannot imagine how early humankind could have survived as hunters. Well, if the early hominids had the same imagination as some women's libbers, then the hunting hominids would not!

Some Hominids Move to Warmer Climes

Humans, Being Black, Have to Get Along

As I have said, humankind evolved in a cool Asia over many millions of years, but Asia has, over the last million years or so, been in an ice age interspersed with many interglacial periods. These interglacial periods may only last for brief periods, compared to the periods of descending ice. In such cases, the habitable area in Asia is somewhat limited. This means that the number of genetic variants in Asia would be limited. However, Africa would be little affected by the Ice Age and would have evolved more types of human beings (take note, DNA people).

During the interglacial, the Asiatic population would have expanded and branched off into many populations. But as soon as the ice returned, the large population of Asiatic humans would be forced Southwards, spilling into the Middle East or South East Asia area and then into Africa, with waves of populations overrunning the indigenous populations and interbreeding with them, please note.

If, at any time, hominids in a cool climate decided or were forced to emigrate to a warmer, sunnier climate, then they would have had to re-develop darker skin in order to resist the ultraviolet radiation. This would have produced the antagonism that existed before. Having adapted to a pack-dependent society, they would have had no choice, in evolutionary terms, but to subdue their aggression (unless under immediate attack), which may have also affected their willingness to hunt. Vegetarianism would, I believe, be the consequence, as would Machiavellianism (social manipulation). I remember my hippy brother telling me many years ago that if one wanted to get a lift (while hitch-hiking), then it was better to wear light-coloured clothes. It appears that people trust or like others better if they wear white (my "hippy" brother did not have race in mind). Also, many psychologists have wondered why fair hair and blue eyes have a calming effect on the human psyche. Is all this just a coincidence? *As I have mentioned before, it may be that one can love family and friends who are black, but not black strangers.*

It is hardly likely that an African population would have left Africa sixty thousand years ago, just when an ice age was occurring, unless they were addicted to chilblains.

Ape Degeneracy

Let us assume for the moment that some evolving hominids in Asia went on a walkabout and ended up in Africa. This may have happened more than six million years ago. As I pointed out above, some humans may well have adapted better to a vegan society. This may well have reduced the pressure to evolve a larger brain and also adapt to the high infection rate in a hot climate, which I have already explained. I would say that this happened to *Australopithecus* and friends about 4½ million years ago. (Although no one knows when the first ape migrations from Asia began, there could have been many apes that migrated to Africa but simply did not survive.) If, a few million years later, another band of hominids decided to "go west," they may well have, by that time, reached a high stage of culture and intellect that they felt no desire to communicate, sexually or otherwise, with the native population. Meeting their poorer relatives in Africa, they would, no doubt, have seen them as competitors. If you have read Robert Ardrey's books about territory, then you already know and understand the following. When a territorial animal sees a competitor, it attacks it on sight, sacrificing food, sex, sleep, and anything else to eject the contender from its territory.

However, prey animals, unless very sick, are only attacked when the hunter is hungry and the prey within striking distance. From a hominid's point of view, competitors walk on two legs; prey, on four. Hominids know on an instinctive level that those who walk on two legs are competitors and that those who walk on four are prey. Therefore, we can conclude the following: two legs = bad; four legs = good. You can therefore appreciate that those lesser hominids that walked bent over, probably on their knuckles, were less likely to be attacked than those that continued to walk erect. Also, ape meat may not have been a particularly favoured part of the hominid's diet unless the hominid was desperate for food. Where extreme human submission is involved, going on all fours is the general rule in many cultures.

I believe that every apparent advance in human evolution in Africa was due to new immigrants from Asia who replaced another group or interbred with it (if the two groups were not too diverse, that is), thereby outbreeding the group that was there first. This probably happened at the beginning of each ice age. As the sheets of ice came down over Asia, so humans migrated away. Biochemists point out that the human species shares a resistance to viruses with the gibbon and orang-utan, while chimpanzees and gorillas have no such resistance. (This has to do with geography, not genetic relationship. Still, it is helpful to know the time of separation of two closely related species. We don't know where the two species were when they separated.) This does point to Asia, not Africa, as the place where humankind originated.

The biochemist who wrote *Monkey Puzzle* pointed out that there are no known fossils of chimpanzees or gorillas. Neither are there any fossil descendants of *Australopithecus.* However, it must be somewhat embarrassing to many anthropologists that the robust *Australopithecus* looks very much like a gorilla and that the graceful *Australopithecus* resembles a chimpanzee. Could it be that present-day African apes "evolved," i.e. degenerated by way of reverse neoteny? That is to say that the apelike characteristics which disappeared with neoteny in the original apeman evolution reappeared when these hominids experienced accelerated development. According to the biochemist, the ancestors of human beings and African apes separated about 4.2 million years ago. Chimps and gorillas separated from one another about the same time. The gorillas went more for the vegetarian way; the chimps, the omnivorous.

During every interglacial, the population in the north of Asia would rise, but during every ice age, the population would be forced southwards and then face both competition and a shortage of land space. This may have led early humans to migrate to Africa, India, and, to a lesser degree, Europe, since Europe was overcome with ice, as well. Each time, Asiatic man tended to interbreed with the natives of these areas. This updated the genes of the tribes in those areas.

However, some hominids stayed in central Asia and suffered the cold of an ice age. This occurred sixty thousand years ago. Judging from the mites that only exist in clothing, this is when human beings first started wearing clothes and probably began using fire regularly. Human beings probably wore hides when on the hunt to disguise themselves as prey animals, as North American Red Indians are known to do. Finding this to be rather warm in a cold climate, they probably made the habit a permanent one. When the Ice Age occurred, they probably made it a permanent fixture. They may well have used hides to make part of their shelters in order to protect themselves from the chill and wind. Cooking may well have occurred when they attempted to defrost frozen food. It is very easy to overdo it when defrosting food, so they may have ended up cooking or burning it. However, those who found cooked meat tasty would also find it to be easily digestible. Therefore, they no longer needed such big stomachs. This means that humankind could therefore become slimmer and achieve better athleticism when on the hunt.

One might wonder how humankind tanned the skins that he had? I wonder if a skin was accidentally dropped in a marsh with a high acidic level and retrieved later and then beaten to soften it up.

Chapter 5

Psychology of Self and Ego

Some terms in this chapter I borrow from Stan Gooch, who wrote *Total Man; Personality and Evolution;* and *The Neanderthal Question.* I would like to point out that I do not necessarily agree with everything Stan Gooch has written; neither would he agree with everything that I have written. But as a psychologist-cum-anthropologist, I would say that he has made a valuable contribution to the understanding of humankind. This is why I borrow some of the terms from his books.

The principle of humankind's dualism, or conflicting personalities in an individual, describes not only a human condition but also an animal one. When life first emerged, a cell was more like a wandering stomach that was only able to collect "food" (complex chemicals) that came its way. This meant that the cell was only "concerned" with its own inner universe. Then, cells evolved an ability to swim, at which point they relied upon odour or chemical detection to home in on food. Then, to put it simply, organisms began to "hunt" for other organisms, thereby relying upon a greater ability to detect things in the outer universe. Of course, the stomach had to feed the cell to keep the system going. One could therefore ask, Is the part of the cell that hunts the slave of the part of the cell that digests, or is it the reverse? Each serves the needs of the other, and neither can survive without the other. When multicellular organisms evolved, some cells specialised in the hunt for food, whether animal or vegetable matter, which, of course, meant that the organisms had to detect light and movement. Other cells specialised in the digesting and processing of food (prey). This is the case for all multicellular organisms on earth today. This does not mean that the self controls the ego when the stomach needs to be fed, but that the self seduces the ego to hunt for food. The feelings of hunger are not mind-controlling; instead, they seduce the ego into getting food, even if the ego goes about it in a roundabout way.

Brain Control

Psychologists have found that different parts of the brain control different functions, as follows. (This is, however, only a rough guide and is by no means absolute.)

Front Left Brain or Higher Masculine
(A) = **Higher Ego**

Analytical verbal thinking:	Analysing the world around
Represents:	Scientist
Approaches:	Abstract, data-based, theoretical
Love:	Knowing that someone is right for you even when there is no emotion involved
Question:	"Do I have all the facts?"
May overlook:	Feelings, synergy (cooperation), opportunities

Hind Left Brain, or Lower Masculine
(B) = **Lower Ego**

Functional thinking:	Being in control, making things work, being a man
Represents:	Engineer
Approaches:	Organised, conservative, procedural
Love:	Sexual desire
Question:	"Will I be in control?"
May overlook:	Alternative solutions, novel ideas, the big picture

Hind Right Brain, or Lower Feminine

(B) = Self

Caring and people-oriented:	This is also to do with the dream state, since many functions of this part of the brain are active even when asleep
Represents:	Mother figure
Approaches:	Interpersonal, intuitive (feelings)
Love:	Falls in love where there is an emotional bond (e.g. an ache in the stomach)
Question:	"How will I affect others?"
May overlook:	Facts, planning

Front Right Brain

(C) = Higher (Self + Ego)?

Fantasy, intuitive:	Non-verbal thinking, depth perception, and visual imagination
Represents:	Artist, writer of fiction, architect, designer, and the like
Approaches:	Imaginative, forward-thinking, risk-taking
Love:	Romantic
Question:	"Have I seen all the hidden possibilities?"
May overlook:	Details, practicality

Evolutionary psychologist Stan Gooch mentions why the brain has evolved in the way it has.

B1 = Autonomic system, which suggests (C), self
B2 = Vascular system, which suggests (B), ego
A = Conscious thinking, which suggests (A) and (D), upper ego

Brain Evolution

First: There is the autonomic system, the part of the brain which controls digestion, the heartbeat, and liver and kidney secretions – in other words, unconscious control. This system cannot detect the outside universe directly. It can only know about the outside universe through its vascular system. This means that the autonomic system lives by having "faith" in the outer universe's existence. Using words that it borrows from the ego, the self judges the world according to its own standards and misuses many words, especially *fighting* and *doing* (more about this later). The autonomic system does not in itself control the hunting (vascular) system. Instead, it seduces, influences, and persuades it to act. When the organism is asleep, the ego is inactive or in a temporary shutdown, but the inner body is still active. Therefore, the self can make itself conscious in the form of dreams. It has also been suggested that in the sleeping state, the brain works to process information and tidy up the input of the day before. This part of the human psyche, and people for whom this system is dominant, are called B1, (C), or self-dominant.

Second: There is the vascular system in the brain which controls muscle movement and power. Those in whom this system is dominant are called B2, (B), or lower ego-dominant. This group includes the hunter, the athlete, the warrior, and others of that ilk. People governed by this system live by and demand proof, as they are in direct communication with the physical universe.

Third: There is the analytical system which evolved from or as an extension of the vascular system. Those in whom this system is dominant are called A, (A), or upper-ego-dominant. The analytical system can be used by the (B) system for logical thinking and by the (C) system for verbal social communication.

Fourth: (D)-dominants are a bit of a mystery. It would appear that this is the area of the brain responsible for visual and three-dimensional perception, fantasy, etc. This system can be used by the (A) system for invention and the (B) system for depth perception while hunting and by the (C) system for symbolic "fantasy" communication. It has also been remarked that the right forebrain has not evolved in the same way that the left forebrain has. This would account for its undisciplined nature.

There are many archetypal images which I believe represent different brain systems. Following are some examples.

A = Scientist, Wizard (wise old man)
B = Barbarian, Warrior ("being a man")
C = Mother (Earth Mother)
D = Artist, Fairy Queen (fantasy)?

When an animal lines up its prey for the kill, the part of the brain that controls the vascular system cuts itself off from the vascular system and rehearses what the vascular system is going to do. It is this behaviour that led to the evolution of consciousness from the controlling part of the vascular system, which is detachable from the vascular system. In other words, this part of the brain preconceives the hunt or any vascular activity planned in advance.

Normal flight from a hunter is not a condition for intelligence. In such a case, the prey has no time to think about or rehearse its actions.

Primates that live in trees need to think ahead and work out, in advance, the best route to take. They also need to decide whether each branch is strong enough to hold their weight or not. It may be that in the beginning, when the first apes came down from the trees, they rehearsed an escape route to safety up in the trees, which a normal herbivore would have no need to do. Being safe means only being able to outrun the hunter, not reaching a safe place.

As far as the distinction between the male and female genders is concerned, I am not saying that women are (C), or **self**, and that men are (B), or **ego**. I am only saying that women tend to be more **self**-dominant than men and that men tend to be more **ego**-dominant than women, relatively speaking. However, I think that as far as the upper brain is concerned, men tend to use the upper left (A) for verbal, analytical thinking, whereas women tend to use the upper left for verbal social communication. Men tend to use the upper right (D) for depth perception when playing sports and when any other physical activity is concerned. (D) is also used for imagination, architecture, and art, and when original inventiveness is concerned. Women tend to use the upper right (D) for fantasy and, perhaps, Machiavellianism with ACD integration, but I am not too sure of this at the moment. Women also use (D) to pick up on subtle social visual signals. I would point out that this is only a rough generalisation, as far as the functional positions of the brain are concerned. The masculine and feminine factors are interchangeable. (The ideas in this chapter, as with the entire book, are not carved in stone.)

The **self** is associated with the left wing of politics, which is matriarchal in nature, demands faith rather than proof, and believes that the inner nature of a person is more important than his or her physical abilities and outward appearance. The **self** is associated with the collective. It does not like divisions of any sort, such as those related to gender, sexual orientation, the family, the nation, race, etc. All the world should be one, and all should owe allegiance to that oneness (cf. the communist state). The **self** is patient like the cat, passively when waiting (lying in wait, a "sleeper") but impatient when in motion (it will keep going only so far before it gives up to do something else). The **self** tends to solve physical problems by taking a short-term view. Also, the **self** perceives truth according to its moral value. If a scientific fact contradicts the **self's** moral value of goodness, then it will perceive the fact as false. To the **self**, moral goodness is truth. The **self** values goodness as the measure of its own survival. Goodness is a measure of one's ability to feel, whereas evil is the absence of feeling.

When arguing with **self**-dominants, one learns that they believe themselves to be right unless they are proven wrong absolutely. In other words, they are "innocent until proven guilty". However, from the **self's** point of view, the **ego** is wrong unless it can prove itself right absolutely, meaning that "the **ego** is guilty until proven innocent".

The **ego** is associated with the right wing of politics, which is patriarchal in nature, demands proof rather than faith (although, when conclusive evidence is unavailable either way, it will settle, for the time being, for what it considers the most likely explanation), and believes that a person's physical abilities are more important than his or her inner nature. The **ego** likes divisions and enjoys seeing everybody in their place. It believes in the family unit, nationalism, and gender, with each sex in its place. "Men do this! Women do that!" Any sexual deviants have their place. Be sure to wave the flag! Each race should know its place, and there should be no interbreeding. The **ego** is impatient like a dog lying in wait, but it is patient to keep on going. The **ego** tends to solve physical problems by thinking of the long-term view. The **ego** tends to value truth, too, if it is factually correct, not according to its moral goodness. To the **ego**, proven scientific facts are truth. The **ego** values goodness as the measure of its own survival. Goodness is measured by one's ability to think and do, whereas evil is the absence of thinking and doing.

Good and Evil

It is often thought that good and evil are two sides of the same coin and that only two choices are possible. This is because to the **self**, the **ego** is evil, and to the **ego**, the **self** is evil.

Well, I would say to the balanced mind, the mind of reason, that goodness is a pathway with a deep chasm on either side of it. One side is extreme **ego**; the other, extreme **self**. Evil is to fall either to one side or the other. Goodness is to walk the pathway without falling into either chasm. Of course, it is very easy to avoid falling into the **ego** by backing very far away from it, as if the farther you back away from it, the safer you are from evil. But then you can so easily fall into the **self** on the opposite side. To many people, the correct pathway may be more to the left or the right, according to one's point of view.

The Good and the Bad about the Self

The **self** has a good subconscious memory. This means that it remembers things that were of no consequence at the time and is able to recall these things when the necessity arrives. The **self** also exhibits wishful thinking, or wish fulfilment. This is a dream or a wish for the "impossible." One normally finds examples of wish fulfilment in fairy tales, which are rife with crystal balls or magic mirrors (ex. TV sets) or magic carpets (cf. cars, aeroplanes, etc. They also show the ability for people to communicate with others who are a distance away. This indicates psychic thought (cf. telephone and radio).

One trouble with the **self** is that it has no desire to achieve or realise its dreams, even if the resources exist to make the dream a reality. No, the **self** only *wishes*. (I would point out here that many people may wish for things but do not have the resources to realise their dreams; therefore, they can only wish. This is not the same thing.) Also with the **self**, there is the desire to seduce and socially manipulate. This may be beneficial to the individual, who may seduce favours from the group, but Machiavellianism does not in itself produce anything useful. More on this later.

The Good and the Bad about the Ego

Analytical thinking, that is to say, a desire to build and construct, often saves work in the future. **Ego**-dominants possess a high degree of physical energy as well as an obsession with that to which they are dedicated. One negative thing about the **ego** is that, unless something or someone proves something possible, it regards that thing as impossible. It does not occur to the **ego** to attempt the "impossible" unless it is seduced by the **self**. I remember hearing a Swedish woman saying on TV, "You English are lucky. You still have something to aim for. We have everything and, therefore, nothing to aim for, nothing to live for." Such a statement would make a balanced person want to weep, as it is obvious to any normal person that one's imagination will always exceed the resources at one's disposal. One cannot judge all Scandinavians on the basis of one woman's remarks, but I do believe that her attitude is not unusual for a Scandinavian. Scandinavians are the purest manifestation of Neolithic man. Need I say more

Result of EGO & SELF

A skeleton that was found in Israel is considered a hybrid of Neolithic man (modern man or Caucasian man) and Neanderthal man. When flesh and skin were added to the skeleton, the result looked not like a typical Arab or Jew but a classical one. I would therefore suggest that the Jewish Arab people are hybrids of Neolithic man and Middle Eastern Neanderthal man. When modern human beings interbred with this type of Neanderthal, they produced a type of human that had a higher ego (i.e. was more intellectual) than Neanderthal man but was more imaginative than Cro-Magnon man and a more integrated self and ego. This gave rise to the true inspirational inventor which, in turn, gave rise to civilisation. Unfortunately, civilisation breeds Machiavellianism faster than intellectualism, especially in warm climates where one can live on minimal food and resources. I have heard in modern Israel the North-Western European Jews do far better than their southern and eastern counterparts, even though they were brought up together in the same nursery. This is because, I suggest, in the north, Jews were in competition with the Gentiles for resources, since, as is the case for Northern Europeans, it is necessary to keep warm in the winter and feeding oneself and one's family. Therefore, one needs more resources in a northern climate, which puts pressure on the selection of the fittest. The most productive Jews who are willing to work in cooperation with the Gentiles and not rely on Machiavellianism would survive the best. I do not claim that my theories should be carved in stone, but merely that this is the general direction in which to look.

Political Correctness

It is generally thought that political correctness (PC) is a philosophy that was created to ease humankind's suffering, an idealistic way of achieving general goodness. I would say that this is not so. PC is governed by the self's deep instinct. Its main aim is to create a oneness of humankind. The only personal bond it recognises is the one between mother and baby. All other bonding between humans is not encouraged. The exception, of course, is the bond of the oneness of humankind.

With PC, pride in one's national identity is discouraged, along with gender-difference awareness and racial identity. Also, the bond between men and women is considered irrelevant, which means that the bond of the family is secondary to the brotherhood and sisterhood of humankind. This is well known in a communist state. While the communist state has lost the economical battle for the mind of humankind, the sociological battle wages on.

Those who espouse PC have done everything imaginable to break up the family, what with hiding behind women's liberation and encouraging schoolgirls to plump for a career rather than to become wives and mothers, which many girls would be happy to do without PC interference. To encourage men and women to live together without marrying or make separation easier. Therefore, there is less reason to be tolerant, as intolerance is bound to devalue marriage.

Consider how much pain and suffering the breakdown of the family has caused. Also consider the number of lives that have been ruined as a result. Then, compare how much suffering has been caused by racism and nationalism in Britain over the last fifty years. I would suggest that PC has a lot to answer for. Take murder as an example. Murder is caused more by intolerance in the family than by any intolerance of race. Perhaps if the PC brigade were to condemn people as ignorant and bigoted for having an unhappy marriage or wanting to divorce, there would be fewer family break-ups and, therefore, fewer one-parent families. Of course, this very suggestion would be an outrage to a person who practises PC, as it is not compatible with the PC belief system.

Use of Words

Words created by the ego may mean different things to the self. For instance, the word *fight* to the ego indicates a man-to-man head-on clash, a competition for a passive prize. Whoever wins the fight wins the prize absolutely. The self, on the other hand, only perceives struggle, conflict, and competitiveness; therefore, anything that smacks of a struggle, conflict, or competition is perceived by the self as a fight. Since the self is told by the ego that fighters are brave, the self therefore considers its own form of competitiveness as courageousness. Even two people competing with one another to submit, crawl, or grovel. The most "seductive" are perceived as "brave" fighters, even though the ego would see the opposite as true. If two women compete for a prize using the basis of who can submit the most, then they only know who has won the prize when someone tells them so. Cf. two prostitutes competing for the same "john".

The word *doing* means different things to the self than it does to the ego. The self – which, by its nature, is a seducer – considers using any form of persuasion and seductiveness in order to get someone else to do something. This is considered "doing." That is to say that the self-dominant person actually takes credit for the job done, even though someone else did the job. This is because, to the self, seducing and persuading are its roles in life.

To understand how people perceive the universe, let's look at stories of ancient times about witches who cast spells on people. The normal thing was for the witch to say, "I will turn you into a toad [or frog or rabbit]." If she made this threat to a man, then he assumed that she would physically turn him into some animal, when, in actual fact, she only meant, "I will psychologically, spiritually, or mentally turn you into a toad [or frog or what have you]." In other words, the witch meant, "I will make you *feel* like a toad or frog, low and unworthy" (or like a rabbit, timid and devoid of confidence). The witch is a classical Machiavellian, so men of the past were only too happy to burn witches at the stake, if they had the courage to do so. I remember hearing that long ago, a witch who was put under arrest for practising witchcraft. She so terrified her persecutors with threats and curses that they let her go. The witches who were burnt were more likely to be innocent, unlike the ones who were released.

There is the classic fairy tale of the prince whom a witch turned into a frog. He could only be turned back into a prince if a princess showed love to him. Also consider *Beauty and the Beast,* wherein a selfish prince is turned into a beast. Only the love of a beautiful woman can turn him back into a prince. To help the reader understand these symbolic stories in the real sense – since all stories have an element of psychological truth to them – I would suggest that each is a case of a plain woman's being rejected by a high-born male, after which point the Machiavellian woman insults and humiliates the man in such a way that he feels low and unworthy. Then, a high-born woman comes along and expresses love for and adoration of him, telling him how wonderful and worthy he is. He is, of course, transformed into the confident, healthy male that he was to begin with. I believe that the reader knows that this happens frequently in the present, as it did in the past. Both the witch and the "beautiful" woman are Machiavellian in their own ways, but the witch seeks to destroy a man's soul, whereas the beautiful young woman wishes only to save it. In times past, a witch, in a real sense, was a woman whom men rejected. Perhaps she was seduced at a young age and then was rejected as though she were only good for one thing, a good screw. At that point, she felt worthless to society because she lived in a man's world. The woman then lived with a hatred for all men and hoped to become powerful and successful, if given a chance in a man's world, so she could destroy, dominate, and humiliate men as she herself was humiliated.

Instinctive Response

While we are on subject of psychology, I would also make an important point about instinct. There is always the controversy over how much of human behaviour is instinctive and how much of it is culturally taught. To make a point, I will discuss the principle of response when a fear of snakes is concerned.

It has been noticed in laboratory monkeys that the young show no fear of snakes to begin with. If a young monkey's mother shows extreme fear of a snake or is attacked by a snake, this will trigger a response in the young monkey, which for evermore will have a fear of snakes. People who live in an area where there are dangerous snakes also have a natural fear of all snakes. When I was a young child, I was bitten by a large dog, although not badly hurt. It still was very frightening experience at the time. However, this did not put me off dogs. I was still happy to play with them. I would suggest that if I had been attacked and hurt by a snake, then I would have developed a phobia of all snakes. However, if I had been badly savaged by that dog, then I would most probably have a fear of dogs, as well. The point I wish to make is that I would need to be hurt very badly by a dog to have the same fear of dogs that I would have of snakes if I had been mildly attacked by a snake.

Sometimes, it is not that we are programmed to fear an animal absolutely. Rather, we are programmed to fear some animals more than others if we are attacked by them. In other words, many responses lay hidden within us. Only a small stimulus in a particular direction leads us to respond in a full-blown way, whether the response be one of fear, love, desire, hate, contempt, revulsion, or something else.

In this book, when I use the terms *instinct* and *natural response,* I mean them in light of the context provided above. Animals have a natural fear of human beings, but those animals that have been brought up with people show no such fear. It is known that some primitive tribes eat grubs and insects with relish, much to the revulsion of Western people. "Why are Western people so revolted?" one might ask. Insects and grubs are nutritious. Could it be that since insects are found so easily, there are no selective pressures on those who eat them and find them delicious to evolve intellectually? Can they therefore afford to be lazy? It is strange that the most people who eat insects are those who live in barren wildernesses, however there are people in the East who might contradict this, but this is the general overall picture. They seem to live on the margins of humanity rather than being a dynamic part of it. The point I would make is that if a Western child were brought up with insect-eating people, then he or she would eat them, too. But it would not take much to put a Western child off of insects if his or her parents expressed a revulsion. Would a child be put off chocolate if his or her parents expressed a revulsion for it? I doubt it. Chimpanzees and many other primates live off of grubs and insects. More on that later.

Chapter 6

Variety of Human Beings in the Past

Palaeolithic man = Neanderthal man (Old Stone Age) – Beetle brow, protruding jaw, thickset, thick-boned, and very strong

Peking man – found in China

Neolithic man = Cro-Magnon man = European Caucasian (New Stone Age) – High forehead, straight face (no protruding of the jaw), slim, athletic

About half a million years ago, there lived a group of humans in Northern China. Their bones were discovered at the beginning of this century. By some "coincidence", the Peking men, as they were labelled, possessed physical characteristics which are known only in the Far East Asian race, such as shovel teeth, a flat thigh bone, and the particular shape of the upper jaw and cheekbones. It is therefore obvious that Peking man is an ancestor of the Far East Asian race as well as neolithic man. Peking man had thick molars with extra enamel around them. These enlarged molars, often called bull teeth, suggest a vegetarian diet. It's believed that Peking men practised cannibalism since many skulls with openings in them – indicating extracted brains – were found, as well. Cannibalism may well have been ritualistic for Peking man rather than something done to add extra protein to the diet. Peking man lived on a large continent at a high latitude, where the weather tends to be very hot in the summer and very cold in the winter. So, Peking man may well have adapted to both extremes of temperature. The fact that the Chinese are known to outwork white people in both extremes of temperature has to do with their Peking man ancestry, I believe.

Although these humans lived in the Stone Age, they did not use fine stone tools, as I understand. It has been suggested that they used bamboo instead. Bamboo can be turned and sharpened into knives. The resin on the outside of the bamboo stalk is very hard and can make an effective knife. It is thought that the Far Eastern man does not have the same inspirational inventiveness as the Western person does, even though the Far Eastern man does not lack intelligence. While it may be a long shot to suggest this, I believe that when a Caucasian cut flint, he had to imagine the grains within the flint in order to shape the tool that he wanted. This required three-dimensional imagination. Perhaps the modern Far Eastern man lacks three-dimensional imagination, which inability he inherited from his Peking man ancestors.

However, the Far Eastern man has a propensity to observe anything unusual and think, *How can I use this to my best advantage?* Silk, rice paper, block printing, are a few examples.

Those who study DNA would say that the Far Eastern man genome matches the genome present in the rest of humankind, but this does not mean that the genetic characteristics are the same, as well. Evolution is based on the selection of manifested genes, not on genomes. Genomes are based on a lot of junk genes that don't do anything anyway. Also, only a small percentage of genes have any effect on an organism.

Rhodesia Man

About two hundred thousand years ago in Africa, there existed a human being called Broken Hill man, or Rhodesia man. Unusual for a Palaeolithic human, Broken Hill man had a rather tall, graceful, athletic physique. His bones were long, stout, and solid, but modern in shape. His teeth were large like all Palaeolithic humans', and his upper jaw protruded out in a straight manner. These characteristics are common, to a lesser degree, to the modern South Saharan African (SSA). However, Broken Hill man's brain was rather small, so it appears that he evolved the body of a man before evolving a large modern brain. This shocked the anthropological field since it meant that humankind developed a hunting and killing lifestyle on more of an instinctive level rather than on an intellectual level. Another specimen of this creature was found called Rhodesia man, meaning that Broken Hill man was not an unusual deviant.

It is a known fact that female chimpanzees are very promiscuous by nature. They mate with several males. The female gorilla, by contrast, is faithful to one male. The male gorilla is polygamous. It is no coincidence that the chimp has larger genitalia than the gorilla. (It is a known zoological fact that the larger the genitalia in the male the more promiscuous the females) The chimpanzee is omnivorous, eating meat at every opportunity. Because of this, chimpanzees are more susceptible to disease, therefore needing to breed as fast as possible with as many different individuals as possible. The gorilla, being totally vegetarian, is not. Africa does not encourage intelligence in a meat-eating species, which is why I do not believe that Africa is the birthplace of humankind.

In hot climates, there are a great deal of viruses, bacteria, and antigens which are continually mutating to counter the adaptability of their hosts' defence systems. The greater the variation of genetic chemistry in the potential host, the more likely the host is to have the ability to fight the antigen. As a result, more young are born. Actually, I should say that the more fertile the population, the better the chance that the tribal population consists of individuals who have an immunity to new strains of antigens. Therefore, so long as there is readily available food, the selective pressure is on fertility rather than on intellect. If a quantum leap in intelligence were to be introduced from outside and if interbreeding were to take place, then not only would there be the benefit of instant boost in intelligence, but also a new gene pool to boost the immune system of the introduced Asian tribe to the indigenous Broken Hill tribe. It is quite obvious that to keep intelligence going once it has evolved is quite different than going through the complex business of evolving it in the first place.

It comes to mind that if humankind originated in Africa, then why would SSA's woolly hair be absent from all the other races of humankind during this short period of time? While woolly hair evolved in Africa, it is not a detriment to humans in other environments. It is unlikely that woolly hair evolved in Africa after humankind's other ancestors left the continent thirty thousand years ago. Also,, Cro-Magnon man, a fully evolved Caucasian, was found living in Europe thirty thousand years ago at the time when human beings are supposed to have left Africa.

Middle Eastern Neanderthal

About fifty thousand years ago, there existed a generalised Neanderthal man in the Middle East. He was a very successful hunter who seemed to hunt by entrapment, getting prey to kill itself by using deceptive tactics such as a fire or great noise – or both – to drive the prey over a cliff, or generally lying in wait in a tree and then jumping on the unsuspecting herbivore. Wrapping his strong arms and legs around the prey in a stranglehold, he waited to be joined by his nearby mates. In this, Neanderthal man relied on luck, but made the best of what luck offered. Relying on luck demands faith, since luck, by definition, is, in itself, out of one's hands. Neanderthal also showed signs of having religious beliefs, such as decorating graves with flowers and performing rituals. One skeleton of a Neanderthal male who lived to old age showed signs of early brain damage and physical degeneracy. It is hardly likely that he was able to hunt. Therefore, he was probably tolerated and looked after by the clan, rather like what is done in a welfare state. Middle Eastern Neanderthal was rather short and robust but not as grotesque-looking as, say, the closely related specialised European Neanderthal. A hybrid skeleton was found in the Middle East of a Neanderthal and Neolithic man. When reconstructed with skin and flesh, it looked, in structure, not like a typical Jew, but like a classical one. Now Arabs may well have evolved from this meeting but Arabs and Jews have gone their separate ways genetically and mixed with different people since, to give themselves a slightly different appearance.

European Neanderthal

There existed in Europe about fifty thousand years ago a specialised Neanderthal of grotesque proportions. He was short, of thick, robust stature, and immensely strong, with a large head and brain capacity. But the shape of the brain, flatter than Neolithic man's, suggests an emphasis on memory rather than intellect. It would appear that he was adapted to his environment on a physical and vascular level rather than on an intellectual level, as was the case for Neolithic man. It seems that he did not leave many descendants and, therefore, had little influence on the European gene pool. The reason why European Neanderthal man did not (from what we can tell) leave many hybrid descendants is because he was so specialised that he appeared totally repugnant to Neolithic man. Also, because he was a very physically powerful individual, Cro-Magnon man saw him as a threat. If Cro-Magnon got caught by surprise, grabbed bodily by this Neanderthal, then he would not have stood much of a chance. Cro-Magnon man would only be able to kill the Neanderthal if backed up by many others who all used superior jabbing and throwing weapons. Therefore, I suggest, as many anthropologists have, that this Neanderthal did not die of measles but was systematically wiped out (which suggests to the "Nature is best" brigade that ethnic cleansing is only *natural,* perhaps. But who says that nature is best? I don't).

Neolithic Man: Where Did He Come From?

It is rather disturbing that neither Neolithic man's nor modern man's homeland has been found. It would appear that Neolithic man was a fully developed Caucasian who came from some place unknown. Most anthropologists still believe that Africa is the place where modern human beings evolved. I doubt that Neolithic man evolved from Broken Hill man since there are traces of Neolithic man in Western Asia at around the same time. It has been suggested that India is the place of human evolution. A large proportion of the population in India is Caucasian. Although there is an element of Neanderthal genes in the population, it is still predominantly Caucasian. When you consider that East Asia, Africa, and America all have non-Caucasian populations, it does seem that Caucasian-ness predominates around India, Western Asia, and Europe. Since Europe and Asia were covered in ice for most of humankind's evolution, I would suggest that the birthplace of humankind and our main line of evolution is in Southern Asia around the Caspian Sea. It is strange that India is the place that produced apes many millions of years ago. Today, however, there are no apes in India other than in the far eastern corner, which is isolated by mountain ranges. Could it be that evolving humankind wiped out other apes in India at the very beginning, refusing to tolerate any kind of competition until their dominance was absolute? It would appear that learned knowledge and a superior intellect were the major survival criteria for Neolithic man, not simply adapting physically to his environment. It has been pointed out that Cro-Magnon man had a slim physique and was therefore a warm-weather creature ill-suited to the cold Ice Age. I would suggest that the Nordic man, whose colouring suits a cold climate, is slim in build and shows a tendency not to sweat as freely as people of other races who live in hot climates. Are some anthropologists suggesting that a race of humans could change colour faster than they could change the shape of their physique? I would point out that Neolithic man was able to survive the Ice Age because he manufactured and wore clothes. Wearing clothes means, of course, that one can keep warm without one's slim, athletic physique changing to a short, round one. Although Eskimos wear clothes and have, at the same time, a short, rounded physique, they live in the very extreme cold of the Arctic. I would suggest that Neolithic man evolved in a more moderate climate similar to Scandinavia's today. I would suggest that Neolithic man was not a slave of his environment but was culturally adapted to meet that environment. It's like this: When a person goes into outer space, he or she adapts not by changing him- or herself, but by wearing a suit that duplicates the earth's atmosphere in side. When human beings put on clothes, they are keeping in the atmosphere of a warm, moderate climate. When a sea diver dives into very cold water he will put on a dry suit which keeps out the cold water. When we build a house with central heating, we are creating our own environment, which approximates the tropics. (Modern people may chose to isolate themselves from what they do not like, rather than change themselves to adapt to a changing conditions, whether those changes are coursed by nature or by politicians).

So if a people do not like anything that, that surrounds oneself "one gets over it" by keeping what one does not like away from oneself. One might believe, rightly or wrongly that one has a right to isolate oneself from unlikeable situations.

Analysis

It has been said that Africa is the origin of humankind because there are more genetic types in Africa than in either Europe or Asia. I will not argue with that for the simple reason that much of humankind's evolution occurred during the Ice Age. This means that for thousands of years in the past, huge ice sheets covered Asia, leaving only the south-western part open for humankind's survival. Being small in area, the south-western section of Asia could only support a small population of human beings at any one time. If many types of humans had appeared during the interglacial when the ice retreated, it would only mean that when the ice returned, they would either die where they stood or be forced to move south and west and then merge with, take over, or be taken over by other tribes – or else move into Europe or Africa and then merge with the foreign races there in order to adapt to the climate which the Caucasians would have been ill-suited to survive. Once interbreeding took place, natural selection would be favourable to those who inherited modern man's intelligence (a large brain, for instance) and dark skin (amongst a variety of characteristics of the native population) to resist skin cancers.

It is quite obvious from the above circumstantial evidence that Caucasian man often migrated from his home base (at the end of each interglacial, as a new ice age crept down from the north) and interbred with Neanderthals in the new area to which he wandered. In a new environment, the Caucasian–Neanderthal hybrids may have had an advantage, as they had the best of both worlds: the quantum leap in intellect from Neolithic man and the environmentally adapted physiological characteristics of the Neanderthal, as well as environmental "intelligence." I believe that these hybrids produced the races of humankind today. I would point out that the original Peking man skeletons which pointed to the semi-Neanderthal origins of the Chinese subsequently disappeared under mysterious circumstances while sitting on the dockside awaiting shipment at the beginning of the Second World War. I suppose it would be far-fetched to suggest that the Chinese authorities saw to it that these "embarrassing" skeletons "accidentally" disappeared. The Far Eastern race has always prided itself on being a superior race, declaring all others to be barbarians. Far Eastern race believe that they are all true human beings and that others are deviant. The Peking man skeletons *might* point to the contrary, if anything. No wonder those skeletons "accidentally" disappeared.

The reason that Northern Europe contains the purest specimens of Neolithic man is because the European Neanderthals were so specialised and so utterly repugnant to modern humans that modern people more than likely exterminated Paleolithic man. "Make war, not love." (It is still possible that some interbreeding took place, but I would suggest that this was only in very rare instances, since modern Europeans show very little signs of Neanderthal in their anatomy.) Also, those Neolithics in the middle East who found *all* Palaeolithics repugnant would not have necessarily settled down with them but may have continued on their way. I am not suggesting that modern human beings interbred with Middle Eastern Neanderthals in a relationship reflecting the brotherhood of humankind. Modern humans' attitude may well have been "friendly" towards Neanderthals so as to get what information they could get, but I doubt that the men would have allowed Neanderthals near their women. When a few hunters came across a tribe of Neanderthals, they may have chatted with them to discover information about the strange land they were in. They also may well have had intercourse with Neanderthal women for the fun of it when such a thing was possible, much the same way as early European explorers had sex with black women. Some Neolithic men may have "gone native," but most would have continued on, perhaps leaving half-caste "bastards" behind. (I use quotation marks to emphasise a very possible attitude of Neolithic man.) At one time in the past, a black woman from an Africa tribe who was "had" by a white man was more valuable to other men of the tribe. This is of course when Black men first met white men and thought they were Gods or spirits. I do not know whether the same was true for Neanderthal man, although Neanderthal man may have been aware of Neolithic man's "intellectual superiority" and thought him to be some kind of god. But who knows? When white men first visited various tribes in the last century and early in this century, the tribal people thought they were gods or spirits. They recovered from that delusion after a short time.

Chapter 7

Neolithic Man

Neolithic Man in the Past

I do not believe that Neolithic man lived in caves, since there were too few caves available at the time. The trouble with a cool climate is that one not only has to hunt for meat but also has to keep warm by producing clothes and creating a sleeping place. This puts double the pressure on a human being's intellectual abilities. The sort of plants that are available to eat raw in cool climates are minimal, as fruit is seasonal. Hunting game was the only reliable way to feed oneself. Of course, Neolithic man would eat whatever roots he could find, although the women probably collected these. But in the wild, roots are not in abundance. Those that were collected were mostly indigestible and needed cooking. How the Neolithics would have cooked them, I do not know. They might have cooked them in large leaves (if available), as some cultures do today. Or perhaps they filled a hole in a rock with water, heated it with hot stones, and then added the roots. The only way to have vegetables in winter would be to store and preserve them in summer, which would have put more strain on the intellect.

Speech

There is always the problem of how speech evolved, given that communication is an important part of human evolution. I would suggest the following. When a hunter was stalking game, he probably used hand and eye signalling, since vocal sounds would have scared the game away. However, when the chase began, vocal communication was more ideal because, in the chase, men would want to keep their eyes on the prey just as a footballer has to keep his eye on the ball. To start with, humans would have made simple sounds like those made by monkeys or apes which give warning of danger and information about food. Monkeys are known to emit different sounds to indicate snakes, leopards, and birds of prey. These sounds are normally associated with an emotional outburst. Either there is danger or there is not. But let's assume that somewhere down the line, some hominids gave a danger signal without any emotional outburst or sense of immediacy, instead pointing into the distance. This would tell the pack that there was a certain type of danger in the distance. I believe that the objectification of emotional sounds was the start of language.

Dancing

Every culture has a form of dance. "What is the purpose of such an energy-consuming activity?" one may ask. Well, I would say that dancing is a form of courtship. The men can show off their acrobatic skills to the women, flaunting how fit and healthy they are. In fighting, there is often a reference to one's showing a sense of rhythm in his or her movements. It is a known fact that fighting requires a good sense of rhythm. Dancing is a harmless way of proving this skill, a platform for a man to show women what a good hunter, provider, and protector he would be. Also with dancing, one shows an ability to move rhythmically with others in unison, proving that one is capable of being part of a team (working with the hunt) and showing a sense of unity with the group, everyone moving together and willing to stand together in adversity.

When men dance with women he and she are showing one another that they can work together as a team.

Music

Music does seem to have an emotional effect on people. As an evolutionist, I have often wondered why this is the case. It has been found that many pieces of music were copied from high-speed bird songs and then reduced in speed to cater to our own musical preferences. Also, it has been found that music matches certain brain patterns and that children's IQs can be raised by listening to classical music (but this may only heighten the child's awareness). I don't understand what bird songs might have to do with human brain patterns. I will have to leave the answer to that question to other minds.

Virginity

Virginity is highly valued in traditional Western and Eastern societies. In the past, there were always more women than men, so a man could "goose around" and still expect a virgin to be available when he decided to marry. The question is why, *traditionally,* virginity was so highly valued in a wife. It could be argued that if a woman had been isolated from men for a time and then married, there would be no confusion about the parentage of her forthcoming children. However, I think it goes a little deeper than that. There is such a thing as sexual bonding. If a woman only knows (in the biblical sense) her husband and therefore has not had sex with any other man, she is more likely to become sexually bonded to her husband. I remember reading that women who remain virgins until they marry are less likely to commit adultery than women who have had past affairs. It may also be the case that a woman who can resist the temptation of having sex before marriage is more likely to resist any temptation to commit adultery after marriage. Given man's natural monogamous state, a wife can have only so many children. Therefore, the husband has to be sure that the few children that he raises are his and not someone else's. Those men who are the most careful and "jealous" about what kind of woman they marry are more likely to sire the most children. Even if it is only a statistical point that virgins make the most faithful wives, it is still true that men will always play the best odds. Since a man in times past could not possibly know whether his wife's children were his, he had to have a great deal of "faith" in her. A man would not have faith in a woman if he could not trust her to be true to him. This is where the term *unfaithful woman* comes into play. A man who has been betrayed has made the wrong choice of wife; therefore, in the eyes of his comrades, he is a fool (a cuckold). It would have been wiser had he remained single and looked to invest in his sister's children (his nieces and nephews) instead of having an unfaithful wife. It is observed that men in promiscuous tribes are known to care for their sisters' children, their nieces and nephews, more than their wives' children. Need I say more? In olden days, when there was a higher mortality rate for boy babies and men who joined the army, women outnumbered men by more than three to one. There were always enough women to goose with, as well as enough virgins to marry. Nowadays, women do not outnumber men. If men want to goose around, then they have to accept that they are not going to marry virgins.

I do sense that romantic feelings are conditional to virginity. (Of course, it is possible for a man to fall in love with a woman without actually being romantically attached to her. This sort of love does not have the same commitment.) The following is my personal view only. In a permissive society, romance seems to have disappeared and been replaced by a certain cynical resignation. It would appear that in order for a man to "feel romantically in love" with a woman, he must believe that the woman is a virgin or, at least, a widow who was a faithful virgin bride. This is true to the stipulations of old-fashioned romance.

While on the subject of women, I will say that most of woman's evolution has been for the purposes of reproduction, child-rearing, and doing tasks that men do not like to do, such as monotonous jobs around the camp-sight. In the beginning (at an early stage of hominid evolution), the female did not take food from the male as though she had an automatic right to it. Instead, she had to seduce the male into giving her food by, I would suggest, using sexual enticements. This would explain the French kiss, (In the ape stage that is) which would be a woman's way of taking the food out of her mate's mouth. When dealing with a highly aggressive male, a female had to be very Machiavellian in her manner towards him. Therefore, a woman is, by her nature, more Machiavellian than a man.

Rape

For our early ancestors, I believe that rape may have been a part of courtship, but this would be a one-off instance per monogamous courting pair. *Rape* is an ugly word and not a nice subject, which makes it all the more important to investigate. I wish, before proceeding, to make it quite clear that I in no way justify rape or offer any excuse whatsoever for any rapist.

There have been reports of virgin women who were kidnapped and raped, and then the men who kidnapped and raped them stayed with them rather than going away. In such cases, the (previously) virgin woman and her rapist find themselves bonded to one another (cf. the Helsinki syndrome). Also, after the rapist has run away, some women, in a traumatised state, have said, "He raped me and just left me there." Well, what did she actually expect him to do? Why should he stay around or take her with him? I am not suggesting that the woman consciously wanted her rapist to stay, as such. But I would suggest that in the past, perhaps when our ancestry was at the ape stage, rape might well have been a form of courtship. A male would see a female "intruding" on his territory and perhaps behave aggressively towards her until physical contact was made. Then, the male would become sexually aroused and rape the woman. Afterwards, the male would hang around, since he had no reason to run. It was his territory, after all. But he would then become bonded to the female, whom he would perhaps "rape" many times (in a traumatised very little resistance would be involved) The female, being somewhat traumatised by the experience, would, after a time, no longer resist, as shame would not exist in a more primitive state being. But she might then become bonded to the male. A woman would not be raped if the male could not run fast enough to catch up with her or overpower her without damaging her (since an injured woman would not be likely to bear children). Therefore, a woman who put up a struggle and ran away as fast as she could was more likely to be raped only by a healthy male. If a man does violence to a woman in order to rape her, this shows that he is weak. Other men generally find him contemptible. I believe that this is why men regard violent rape as far more serious than rape committed without violence (at one time, non-violent rape was thought of as somewhat of a joke). I know it's quite provocative to say this, but I believe that the shame which follows rape arises because of the following: No woman likes to be used as a thing for a man's temporary pleasure and thereafter be regarded as unworthy of anything else, i.e. marriage, a permanent bond. This is especially true when she is forced to participate in sexual intercourse. Therefore, a woman's pride is the thing that provokes a degree of shame. If a raped woman is married, then she would not want to shame her husband and raise any confusion about whether any child conceived was his or not. This does not take into account the emotional upheaval involved when she considers whether she fought hard enough against the rapist or that she may not have been careful enough and had thereby put herself into a precarious situation with a man who aimed to rape her.

If a woman is married to a weak man or one she thinks of as a loser, then she may prefer to mate and have children with a strong man. It would be in her genetic interest to be raped by a stronger man, so long as there was not too much violence. The woman may place herself in a vulnerable position and let a strong man "rape" her. This would relieve her of culpability when she was expecting his, not her husband's, baby. However, the rapist could then say, "Oh well, she was asking for it."

Rapists often say this. How does society or the husband know whether the woman wanted to be raped or not? In fact, the rapist could, given cultural misinformation and his wishful thinking, believe that she wanted to be raped, no matter how much she struggled. Unfortunately, it's all a matter of faith. A husband may be confident that his wife did not want to be raped, but this does not mean that everybody else thinks the same. In this case, the man is left looking like a fool in the eyes of society. His own wife sees him as a fool and a loser – and she wanted to have sex with another man. How does a wife disprove such an allegation? What can she do other than protest? (Unfortunately, some women have taken their own lives in order to protect the honour of their husbands and families. "Taking my life proves that I had no motive to allow myself to be raped.") If a man rejects his raped wife with an instinctive feeling that she is no longer his, then he has a reduced chance of being cuckolded by the rapist and is therefore more likely to produce children with a different wife. Such a thing probably happened in the past as a matter of course. However, for a modern man, who has a greater social awareness although he may retain primitive feelings, there would be a great deal of heart-searching, since he would be aware that he rejected his wife for something that was not her fault. For that matter, he may feel (quite unjustly, I might add) that he failed to protect her. It is a known fact that a husband feels as much a victim of rape as his wife. It may be that a husband feels that the rapist degraded him, making him look like a weak man.

If, before she was raped, a woman was a virgin, she doesn't feel shame because a man wished to take her but because she knows that the rapist only used her for a "cheap thrill" and then rejected her after the act, without the act's having any other result. A virgin woman who knew that her rapist, a man whom she liked in the first place, would come to love her instead of leaving her afterwards, *perhaps*, so long as violence was not involved, she would not regard her being raped as a horror. If women have rape fantasies, it's probably in this context. (Not being a woman, I can't be sure of this, since this is a subjective, not an objective, matter.)

Long Hair

A woman with long hair does tend to excite men. When playfully chasing a woman, most men are strongly tempted to grab hold of her flowing hair. As women get older and more mature, they tend to cut their hair or tie it in a bun. In many cultures, women tie up their hair and keep it covered by day, only letting it down at night when they are with their husbands. (This is the traditional fashion in Spain.) I would suggest that when a woman covers her hair or cuts it short, she gives a strong signal that she no longer wishes to be seen as a sexual object. Instead, she presents herself as totally unavailable and not a sexually available person who must now be taken seriously. I would say that a man's grabbing a woman by the hair played a great part in courtship rape. Women have very strong hair roots at the back of the head, roots stronger than necessary for any reason other than this one (unless you think it was for the purposes of performing a circus act). As human beings developed and compatibility became more important to tribal societies (along with the ever-increasing brain size), true rape turned to mock rape. During most courtships and especially after marriage in modern times, sexual intercourse involves a form of mock rape, which the woman may find just as exciting as the man does.

Baldness

Bald-headedness seems to be a Neolithic characteristic. Baldness is not only embarrassment to many a man but also a puzzle to many anthropologists. It is a known fact that humans lose most of their body heat through the top of the head. We are told in extremely cold weather to keep our heads covered. I would suggest that the bald head acted as a radiator to cool a hunter's body in a cool climate so that he did not work up a sweat, which could have caused hypothermia if his clothes got wet. A similar principle applies to dogs. Nordic peoples do not sweat as easily as people of other races do. When man the hunter tracked his quarry, he started off on a steady walk with, I suggest, a hood up to cover his head. When it was necessary to run down his wounded prey, he pushed back the hood to expose his bald head and get rid of any surplus heat. (It is a known fact that dogs and wolves do not sweat through the skin but pant and release heat through the lungs and, of course, the tongue. This means that they do not make their under-fur wet, which would dispel heat and cause hypothermia.) It is also possible that a bare patch developed on men's heads which may well have helped absorb vitamin D from sunlight. Since sunlight is in short supply during the winter months and since humans' skin is covered by clothes, this would have made bald-headedness even more necessary. I would suggest that baldness is a recent evolutionary adaptation. Most women do not find baldness physically attractive, even though it is a beneficial survival factor.

Milk

As a side note, I'll mention that Northern European adults are able to digest cow's milk (nine out of ten Northern Europeans, and one in twenty non–Northern Europeans), but people of other races, such as Far Eastern Man, South Saharan Africanes, and other non–Northern Europeans, find it difficult to digest cow's milk as adults. Desmond Morris said on a TV programme that he could not understand why this was the case. I believe that Mr Morris was being politically correct, since digesting cow's milk is a neotenous characteristic (which, I am sure, Mr Morris is perfectly aware of). Europeans are able to digest cow's milk because they are more neotenous than other races and, therefore, more Neolithic.

I would also like to point out that some doctors query why babies fed on human milk show higher intelligence than those fed on cow's milk. I remember reading about the properties of human milk many years ago in the book *The Hunting Hypothesis* by Robert Ardrey. A Doctor Crawford found that there were two kinds of fats, visible and invisible. The visible fats are what we store in our bodies, on our thighs and stomachs. These are the fats we are warned about, the types we find in marbled meat. There are also invisible structured fats, sometimes called fatty acids, which congregate in the brain and are essential to the normal growth of cells.

That is where the carnivore comes in. He profits by the herbivore's season of storage with a single kill and a single meal. Furthermore, from the simple chains of the browsing or grazing animals, it can produce as a next step in the food chain the building blocks of the long complex fats that our neural and vascular systems demand. No large-brained being need be exclusively a meat-eater since a large brain can still be made if enough time is aloud.

Animal Domestication

The Dog

It has been said that the dog is the first animal domesticated by humankind. While there is no proof, as such, of this, I will go along with the idea, given that the dog can help human hunters. Some documentaries have shown that certain tribal peoples use the dog to clean up a human baby's excrement, but this is only an odd exception to the general rule that dogs are used by human beings only for male-oriented jobs. In recent times, we have seen female centralism emerge, as though all things revolve around the female. The documentary explanation of the origins of dog domestication is simply a symptom of our times. The dog, or, in the beginning, the wolf, probably followed male hunters as a scavenger, since early man would not have carried every bit of the animal, especially the offal, all the way back to a camp site. Therefore, when early man hunted large animals such as mammoths, woolly rhinoceroses, elk, or deer, he would leave plenty of pickings for the scavengers. I do not think it is a stretch to imagine that wolves instinctively knew that the sooner the men made a kill, the sooner they would feed. (In the Arctic, the Arctic Fox is known to follow the polar bear, knowing there are rich pickings when the bear makes a kill.) therefore, it would instinctively be in the wolf's interest to distract the game by baiting or frightening it and leading it towards the hunting party. The men, realising the virtue of these wolves, would be only too glad to throw some meat their way as a reward. It has been observed in nature, in rare instances, that two different species sometimes gang together for their mutual benefit. For instance, a bird in Africa known as the greater honeyguide finds a bees' nest and makes a great fuss, fluttering around to attract any interested party, such as a human or a honey badger. It then waits around for the easy pickings when the interested party destroys the bees' nest and departs. Men are known to leave some honey on a branch for the honeyguide afterwards.

Many young wolves would have grown up in close proximity to human beings and therefore thought of them with greater familiarity, as a part of the pack. Those wolves that were tamest towards humans would, I suggest, be thrown the most meat after a successful hunt. The men who had a natural liking for wolves would understand them better and perhaps adopt young orphaned wolves, which would become socialised, serving as the parent and raising the pups. Eventually, these wolves evolved into the modern domesticated dog. People today who have adopted wolves have found them to be very affectionate and loving towards humans. It is not hard to imagine that wolves were domesticated over many years, when men allowed their best dogs (according to the rules of evolution) to breed and create future generations. It was stated on a TV documentary that dogs were relevant to the evolution of humankind. It would be true to say that human beings have an inferior sense of smell and that their hearing is nowhere near as efficient as most animals'. However, neoteny, the slowing up of the development of mature characteristics, would impair these senses even more, making human beings very vulnerable to sudden attack from either animals or other tribes with better-endowed senses if not the same intellectual ability. These dogs, kept as "pets," would soon learn to give warnings in the form of a puppy "yap." In a large dog, this would develop into a bark. The greatest use of the domesticated dog is its ability to give warning, which dogs are only too ready to do. This would be very useful when Neolithic men faces conflicts with Neanderthal men. From then on, it was only a matter of time before humankind used dogs for a variety of uses. Those dogs that did a particular job best would be allowed to breed.

Horses

How the horse became domesticated is a bit of a mystery. I would suggest that a foal was bought home from the hunt or perhaps followed hunters home after they had killed its mother. The foal may then have been kept alive to provide meat in future. But by then, children would have adopted the horse and become attached to it. It's also quite possible that a man may well have ambushed a horse by jumping on its back and putting his arms around its neck, with the intention of cutting its throat with a stone knife. However, if, by some chance, the man dropped his knife or was not at first able to kill the horse, then it's very possible that the horse would have bolted and that the man, perhaps with some apprehension, would have clung on for dear life as he was carried away for a joyride. At some time when he fell off, he might, on the one hand, have been relieved to be alive. But he would then remember the experience of running like the wind and recall how smooth the ride was when the horse was not bucking. When I rode a horse for the first time in Spain, the instructor commanded the horse I was riding to gallop, which frightened the life out of me. But after pulling the horse up and recovering from the shock, I remembered how smooth the ride was. I then voluntarily spurred the horse on to gallop, much to my enjoyment. Now, let's suppose that a hunter had a similar experience and attempted to ride a foal that the children in his tribe had adopted. Perhaps, even, a child got on a foal's back and the foal showed little fear. I know this is a long way from horse riding as we know it today. In any event, while a horse can be ridden, humans may have thought it safer at first to have horses pull a chariot of some kind or serve as beasts of burden.

Sheep and Cattle

I would suggest that domesticated sheep and cattle would have started off as newborn young brought home from the hunt and stored to kill later for meat, but some of these became bonded to the tribes people. It would be far more efficient to keep and protect captured animals than to take on the problems of hunting and competing with every other carnivore. Those animals that responded by staying close to humans and remaining with the herd instead of straying would, of course, continue to breed in captivity, their offspring inheriting the genetic characteristics that made them liable to remain with human captors. A similar principle would apply to those animals that were the least aggressive.

Chapter 8

Analysing the Different Races of Human Beings Today

Caucasian = Neolithic man = modern man = native European

If it is to be assumed that all non-Caucasian characteristics originated from the Neanderthals of that particular area that the characteristics are found in. Example The Far East Asian race has characteristics that existed in the ancient people of that area. I would suggest the following. I would point out, however, that the characteristics I refer to indicate the purest forms of that race and do not necessarily refer to all the people who belonged to said race. I would also add that a race is defined as group of organisms that have characteristics in common with one another and that are distinct from other groups of organisms that also have characteristics also in common with one another. The two groups are, however, quite capable of interbreeding.

On East Asian Race (Chinese and Japanese)

The Far East Asian are unusual in that they have extra-long stomachs, as one would expect to find in herbivores. This gives their skin a sallow colour. Peking man, who had bull teeth, as I mentioned before, is often associated with an herbivorous lifestyle. I would say that this was a characteristic of Peking man and his descendants before interbreeding with Neolithic man. Therefore, Peking man and modern Far East Asian Race are natural herbivores. Of course, it would have been necessary for Peking man to defend himself against wild animals and people of competing tribes. At one moment, he might be calmly eating vegetables; at the next moment, he would have to shift into a state of fight or flight, which is a high vascular state (B2, ego).

The modern Far East Asian Race does seem to be obsessed with the coming together, or integration, of yin and yang. Is this concept the equivalent of the Western concept of self and ego? I believe that this obsession does not reflect what the Far East Asian Race have, but, rather, reflects a way of compensating for what they lack: mental integration. I, of course, mean this in relative terms. A Far East Asian Race's profile may be ABC, but his A and C may not be integrated as well as in people of other races. I am not carving this idea in stone. I could be wrong. Still, I do believe that the brain of the Far East Asian Race is slightly different from the Caucasian's. In fact, I have heard Far East Asian Race (in this case, Japanese people) state this very fact. Of course, it is quite obvious that culture has an overwhelming effect upon the individual. If a Far East Asian Race were to be brought up in a Britain, he would certainly have the mannerisms of a Brit. In fact, one would not be able to tell the difference between him and another Brit unless you looked at him from a racial perspective. This does not alter the subtle differences that are disguised by one's natural conformity, which is done under peer pressure. In the long term, a people will turn to what is in their nature, something which will, after many generations, appear naturally in their culture unless the characteristic or inclination is harmful to their society. More about this later.

The Japanese Myth

The Japanese seem to have their economy all wrapped up, having a splendid reputation for production. The Japanese electronics industry expands by leaps and bounds. I have read, however, that 69 per cent of all ideas that come out of Japan originate in Britain. Only 6 per cent originate from Japan itself. I am not convinced that the Japanese possess superior intelligence. Japanese children do well at school, performing far better on examinations than English children. Their scores are 30 per cent higher, I believe. It is quite obvious that Japanese children are not taught the same way as English children are. The Japanese concept of child discipline is somewhat bizarre in that the children seriously study virtually all the time they are in school. If I and many of my contemporaries had studied as hard as the Japanese children at school and not mucked around as we did, we would not do 30 per cent better, but more like 200 to 300 per cent better, on examinations. Although, to be fair, I note that the Japanese are lumbered with a writing system of "hieroglyphics," which would, I suppose, slow them up. If the Japanese are so practical, as has been proposed, then wouldn't it be a matter of common sense for them to have changed the writing system to something better years ago? (The British educational system has been cursed with teachers trying to teach a generation to read by presenting writing as hieroglyphics – pictures – rather than using phonetics. No wonder the British educational system has been in a mess. At least we Brits are willing to learn from our mistakes and put things right.) I honestly believe that Japanese people's educational intensity is more to compensate for what they lack naturally. In educating children this way, they destroy any natural imagination that children have for true inspirational invention. The Japanese seem to excel mostly at marketing, in knowing what people want to buy, and also at getting people to work efficiently. Getting people to work harder and more efficiently is not only used by the Japanese in Japan but also in countries such as Great Britain and America where the Japanese have taken over.

This suggests that the thing at the heart of Japan's success is not that Japanese workers are cleverer at putting things together. The Japanese tend to be a people-centred race. Their success with the objective, however, is realised by "Borrowing" objective ideas from other nations and then applying basic engineering, mathematical skills, and continuous trial and error to perfect the product. They then use subjective skills when marketing, putting these ideas to their best use.

Also, the people who are part of the huge research and development industry in Japan have been able to perfect many ideas. The Japanese employ about sixty engineers just to stop a car engine from vibrating. In Britain, one would be lucky to find sixty engineers in any one R & D department. I would suggest that much of the Japanese people's success is realised by their ability to use trial and error along with mathematical skills, but they don't necessarily use a great deal of spatial imagination or lateral thinking. Another virtue of the Japanese is their ability to observe and concentrate on details. Many European engineers who have worked with Japanese engineers have noticed this trait. Japanese engineers will study an object's every detail and will not leave it alone until it is as perfect as it can be. Normally, such a process would bore a westerner. Also, if you analyse the average standard of living for many Japanese people, you find that many still live in the "Middle Ages," as they have outdoor toilets and no proper sewerage. (This may have changed recently.) Those on top were only able to build up the Japanese economy by paying low wages and leaving people to go without the benefits of their economic output.(Again this may have changed recently). All profits were reinvested in research and development and automation, thereby reducing the price of the goods produced. If you have enough research laboratories, then you are bound to come up with results at some time. How can any economy compete with Japan's extensive research capability? Although it is said that the Japanese are paid more than the British, this does not mean that the Japanese have a higher standard of living than the British, since a wage is only as good as what one can buy with it.

On South Saharan Africans

I would say, judging from the remains of Broken Hill man (found in South Africa), that, being highly athletic, he was a capable hunter.

Broken Hill man was the type of hunter who relied on the chase and not on seduction to the extent that other Neanderthals did. Considering Broken Hill man's size and brain shape, it seems that he was not an intellectual. However, he may have been bodily clever, that is to say, a highly skilled gymnast who had the ability to think fast and instinctively in the process of hunting. The hunt may have been relatively easy and may have demanded no more than physical fitness, considering the abundance of game in Africa. The main thing about early humans in Africa is that a slight increase in intelligence, for a primitive culture, did not mean greater survival, given rampant diseases. The cause of disease can only be understood by a technologically advanced culture that has microscopes and so forth. Disease resistance, a high breeding rate, and a faster maturing rate are more important when trying to cope with a high mortality rate. Broken Hill man's smaller brain at birth would present fewer problems. However, if a quantum leap in intelligence and culture were introduced to this population, then it would have made some difference. I would point out that since the climate of Africa does not put the same pressures on intellect as a cooler climate does, the Broken Hill man did not have a personality directed towards analytical matters, (I will talk more on this later). I would suggest that Broken Hill man was vascular-dominant or B- or B2-dominant. This is reflected in the South Saharan African's athletic skills, of which we are all too aware on the Olympic field, in the boxing ring, and on the dance floor. I once heard a black man say, "Let's have a laugh and go watch the white boys try to dance." I see his point. I believe (as a pure generalisation, I would add) that basic engineering is a skill at which the South Saharan African would excel if he or she were given the chance. I would also add that the South Saharan African has been subject to natural selection, or survival of the fittest, in recent times, whereas the European has been weakened by living within a protective civilisation for a few thousand years – which means that many Europeans have inherited weaknesses (in other words, they have become genetically impoverished). While the best European may be as athletic as the South Saharan African, it may be that more South Saharan Africans, because they are part of a minority group in Europe, may choose sports, athletics, or some form of show business as a means of getting out of the ghetto. South Saharan Africans tended not to participate in sports such as horse riding and tennis until recently. I believe that this is because these sports are somewhat expensive to participate in, not that the South Saharan African lacks the ability to do well. Also, white people have resisted allowing promising South Saharan African individuals to participate in certain sports, exaggerating the need for intellectual abilities. No player of a fast-moving sport needs intellectual abilities, since intellectual abilities generally require slow, methodical thinking over a period of time, not fast, instinctive reactions on the spur of the moment, which is something that is developed over long periods of training. I would not dispute the South Saharan African's ability for fast instinctive thinking. South Saharan Africans should be afforded every chance to prove their abilities in all forms of sport. As an aside, I do think that everybody should have the chance to prove themselves in sport or any other occupation, not be denied on the basis of dogma. Just because I make points of generalisation in order to explain humankind's evolution and illuminate why different civilisations turned out the way they did does not mean that any individual should be barred from participating in any occupation if he or she has the wish and talent to participate.

I believe that pure, fair-skinned Cro-Magnon man would not have immigrated to Africa. I believe that some Middle Eastern hybrids may have gone south, following the Nile until reaching Central Africa. This may not have been so much an expedition, but more a way to transfer genes. The hybrid tribe may well have interbred with a Neanderthal tribe nearby, which then interbred with the next Neanderthal tribe, following the Nile as they went. Of course, it could be that a tribe of hybrids may well have wandered into Africa simply by expanding their territory by ten to a hundred miles per generation. So, the modern South Saharan African is made up of Cro-Magnon man, Middle Eastern Neanderthal man, and Broken Hill man. Biochemists state that the South Saharan African has the widest array of human genes and that, therefore, all of humankind evolved from Africans. I do not agree, of course. Perhaps the South Saharan African is made up of three original races, whereas the other races of human beings are made up of one or two original races (see the above explanation). The South Saharan African profile may be B or B2. I recall a biochemist stating that the South Saharan African genes have been isolated longer than those of any other race and that, therefore, all humans evolved in Africa. I do not believe this. How long certain races were separated is indicative of their geographically isolation. It is not an indication of geographical origin.

On North American Red Indians

North American Red Indians are a mixture of Far East Asian Race and Caucasian races, but they emigrated from Asia by way of the Bering Strait before they became a consistent and uniform race of people. (There is also a strong possibility that Neolithic man or Caucasians from Europe at the close of the Ice Age crossed over to the American continent by travelling from ice flow to ice flow [as has been suggested by a TV documentary], eventually interbreeding with native Far East Asian Race.

These were the first people to enter America, where the game was innocent of humankind, had not learnt to fear people, and had not acquired what is known as flight distance. In Southern Asia (India) and Africa, the game were not innocent of human beings; they knew them to be dangerous animals. This was not the case in America and Australia. When humans entered America and Australia, the animals did not run away, so they were easy game for the early hunters. Early hunters wiped out most of the prey species as well as many carnivores that presented competition when hunting for game. Ironically, the horse evolved in America. Some horses travelled to Asia via the Bering Strait. The stop-at-home American horses were completely wiped out by the Red Indians. It has been suggested that the horse died out when the climate changed, but the horse is a highly adaptable animal, one that can survive both the cold of Canada and the dry heat of a Mexican desert. Present-day Red Indians, who sometimes claim themselves to be the guardians of the land and mention how much reverence they have for it, are perhaps not representative of their ancestors. The funny thing about the Red Indians, with their sanctimonious attitudes towards the "wicked white man," is that when trading occurred, they did not appear (according to the Hollywood films) to have any interest in carpentry tools or building equipment, but only in guns and knives used for hunting and killing or else vain and frivolous things such as mirrors and bowler hats, which are hardly suitable for the open range. Therefore, I am a bit sceptical of all ethnic sanctimonious talk about what the "wicked white man" has done.

The Australian Aborigine

The Australian Aborigine is a hybrid of neolithic man (Crow Magnon man) and Indian Neanderthal (who was, perhaps, related to the generalised European Neanderthal) that did not immigrate north into Asia with the ancestors of the Caucasians, but later immigrated in waves Eastwards to Indochina, down the peninsula to Australia. What happened to the game in America also happened to the game in Australia, for exactly the same reason. The number of surviving species in Australia is no thanks to the Abo, even though the Abo boast of their belief in taking care of the land. They accuse white people of destroying the land, or at not taking care it, but ever since white people arrived in Australia, there has been a greening of the land. Also, many adaptable species like the kangaroo have actually expanded in population. The number of species that white people have killed off is minimal compared to the number killed off by the Abos when they immigrated, especially as far as the larger mammal species are concerned. The Abo often refer to the Dreamtime. I believe that this is a cultural memory of the Neanderthal before the Neanderthal interbred with the Neolithic hybrid, since the Neanderthal is believed to be a self-dominant. The self, as I have already suggested, has to do with the dream state. It was interbreeding with Northern Caucasians who immigrated south into India that created the East Indians of today.

Harmony with Nature

Regardless of all the propaganda about ethnic people's being nearer to nature and understanding it better than white people, if you look back into history, you will find many instances, from the Pacific Islands to the Hebrides, and from Central America to China, where people have destroyed the land through overuse, in many instances cutting down trees until only peat bogs or desert remained. I honestly believe that when many ethnic peoples see modern white people chop down an area of trees here and there and ruining the land, they simply jump on the bandwagon of "white Western protest" with a sanctimonious attitude, as though they know best and have special knowledge that is far greater than the Western person's. If an ethnic person is the first to protest, it's generally because he has a personal interest, as if some destructive industry is taking away "his" land or scaring off "his" game. The irony is that if westerners pay the ethnic people enough money, then they are quite willing to wipe out a species, even though it is not in their interest in the long term. Remember, it's Westerners who head the fight to save the elephant and rhinoceros, whereas ethnic people generally do the killing. In the past, wherever human beings have gone into new areas where the animals are innocent of them, it's generally "curtains" for any creature that has nice fur or tasty meat. Only in recent times have Westerners not hunted the game to extinction after finding new islands in recent times. Also, it is mostly Westerners who have put so much into saving many species from extinction, even giant pandas, which seem reluctant to survive. When comparing Westerners' treatment of animals to ethnic peoples', I find that Westerners come out as the most gentle, in spite of Western food factories. It is also modern Westerners, not ethnic peoples, who make use of science and analyse the soil, trees, and wildlife with the aim in mind of keeping everything in balance. Western people learn from their mistakes and do not "give up the ghost" of progress. They simply try and try again to make right the blunders of the past. Some ethnics are, of course, exceptions to this rule. However, where ethnic peoples have managed the land after a long history of hard-won lessons, they do not make any progress, make changes, or experiment after a balance has been achieved.

Caucasian = Neolithic man = Cro-Magnon man
Modern man = Native European

No one seems to know where Cro-Magnon man popped up. Some say from Africa; others, from Asia or even India. I am convinced, however, that Cro-Magnon man came from Southern Asia or north of India, which is where we should look to find evidence of modern humankind's origins. I would like to mention again that although India had a large proportion of apes millions of years ago, no indigenous species live there today, apart from a species of gibbon in the far north-eastern corner. Did evolving hominids wipe out the competition? Or, perhaps, when the Ice Age occurred, India became so cool that only apes already adapted to the hominid way of hunting were able to survive, as the Ice Age brought cold winters without any fruit on the trees and no abundant vegetation for year-round food supply. During cold winters, one can live off one's body fat if it is substantial, given that a large animal does not lose much energy when keeping warm. Because of a smaller ratio of skin surface to body weight, a large animal can afford to carry thicker insulation, which a smaller animal cannot do. An animal could, of course, hibernate, but this is not normal for primates. Or, an animal could be carnivorous. Which option would an ape select? I do not believe that our ancestors could have stored enough fat to last them through winter. It is also highly unlikely they hibernated. I believe that hunting apes could only have survived the Ice Age in Southern Asia or India.

The unusual thing about modern man compared to Neanderthal man is his very high forehead and the higher capacity of his brain (for intellectual purposes). Therefore, modern human beings are classified under system "A" or show (A) and (D) dominance. What selective value would such a large, delicate brain have for a human being in a primitive

Stone Age society? I would suggest that Neolithic man evolved in a cool climate rather like that of Northern Europe and therefore needed clothes to keep himself warm. There would be strong survival purposes for producing clothes that not only allowed unrestricted freedom of movement during the hunt but that also prevented air from flowing through. This means that Neolithic man tanned animal skins before shaping and then sewing the leather together. As any tailor can tell you, shaping and sewing cloth, hide, leather, or another material is not an easy job. It takes a lot of doing. And because people come in different shapes and sizes, no one pattern will fit everybody. Also, Neolithic man was very highly skilled at tool-making, which was essential for hunting as well as tailoring. Neolithic people must have many things useful for their survival, such as protective defences and the huts or wigwams they slept in. If you have ever seen a modern furrier work on animal skins, then you know that the furrier does not simply get hold of a whole skin and then shape an article of clothing from it. Instead, the furrier cuts the skin into pieces and then shapes the article from the pieces. I believe it is possible that this is how Cro-Magnon man made clothes. I know this is highly circumstantial, but in our modern world, as in the ancient world, ideas are born from the principle of dissect and unite. Science is a way of perceiving the universe. It cuts everything up into the smallest parts in order to analyse those parts in isolation, and then it joins the separate parts together, according to a scientist's own will. The first example I will use is the phonetician alphabet. Before the phonetician alphabet was developed, human beings "wrote" by drawing pictures. One word was expressed with one picture. Every time a new word was uttered, someone had to create a new picture to represent it. This required a great deal of specialised knowledge, so only a few people had the type of memory required for the task, along with the time to learn and study. However, in using the phonetic way of splitting up words into sounds, one has to remember only about twenty or thirty signs in order to write any word – even words one has never heard before and doesn't even understand.

The science of chemistry was not fully understood until the periodic table was developed. Nowadays, if you want to understand the behaviour of a chemical, you split it into its separate parts and then rejoin those parts to achieve a full understanding of the properties. A significant apparatus used by nuclear physicists is the cyclotron, or atom cracker, which splits atoms into pieces so an observer can come to understand the atomic structure. In carpentry, a person chops down a tree, slices it into lengths, and then builds an article, even a boat, from those lengths. This is a long way from the dugout canoe. Television pictures are created by splitting an image into lines or pixels and then transmitting the data to a receiver (e.g. a TV set), which then joins together the information received.

Many years ago, I read a theory about Cro-Magnon man. The fact is that the Cro-Magnon race remained fit and healthy for a long period of time. They remained tall and athletic in build (like a large modern white American) for many thousands of years. Throughout history, a pattern of events has been observed. When tribes of human beings enter a new territory, they expand their numbers when they have plenty of food and resources, which allow individuals to reach an optimum height and become physically fit. This occurred in America between 2 and 3 hundred years ago. With America's plenty resources belonging to no one except those who can exploit them, individuals have only to compete with other individuals. Therefore, the biggest and fittest win out in the end. Added to this there is also hybrid vigour where many European immigrants of different nations came together, which in itself can produce greater size and health, However, as a population expands, whether by more immigration or by breeding, Survival may no longer be a question of individual physical fitness but a matter of how many there are to gang up on others. Therefore, the Machiavellians tend to take over in the end. After the fall of the Roman Empire, there was a great movement amongst different barbaric tribes, which not only expanded outwards in conquest but also increased in size. This did not last long, however. As civilisation became more widespread and resources per head of population decreased, so did the average size of the individual.

Where a short supply of food exists, there would be a survival factor in being small and short, since a smaller, shorter person would need less food to remain healthy, so long as one had the protection of civilisation to stop you bigger competitors from bullying smaller people.

Of course, an undernourished child would develop into a shorter adult than it would otherwise be if it were well fed as a child. In modern times, technological advances in farming have served to increase the amount of food per head of population. This combined with interbreeding between people of towns and cities produced a vigorous hybrid, thereby enabling human beings to increase in size (but only because of environmental reasons, not genetic selection). This, I believe, had more to do with the womb environment of much healthier women, since a larger and healthier womb creates a bigger and better-developed baby. The females of successive generations inherited healthier wombs, in turn. It has been suggested that Cro-Magnons remained physically fit many thousands of years. Could it be that Cro-Magnon man kept his numbers down and therefore did not out breed his food supply or the natural resources that were available? This would mean that he would have been able to maintain a high degree of health, thereby giving his genes the greatest potential for development. It has also been suggested that Cro-Magnon practised selection upon himself, thereby taking evolution's place and breeding his descendants for Neolithic "perfection." (As I shall explain further on in this book, Cro-Magnon man neither suffered the wastage that comes with having Machiavellians in the tribe nor "perhaps" suffered fools gladly.)

An unusual thing about Neolithic man was his chin, which is something that Neanderthal man lacked. As, during the process of neoteny, the jawbone of Neolithic man became smaller and therefore weaker, there evolved a selective advantage in the front part of the lower jaw. It was turned vertically upwards to give strength to the bottom teeth. When you look at modern bridges, you find a reduction in materials but an increase in strength as the bridges become more sophisticated in design (although they may be more vulnerable to knocks and collisions). This in the principle behind Neolithic man's bone structure, in both his body and his face.

In Cro-Magnon Society the genes of men and women of high rank would choose spouses with characteristics associated with intelligence and physical fitness would survive the best. Examples: A man with a strong sexual desire for a woman who had a high forehead, a gentle, refined, childlike face, wide hips, a slim waist, and large breasts would choose her as a mate because he would more likely to have healthy children with her. There is a high probability that the woman's pelvis would be the right shape for giving birth to a large-headed infant. Any fat may disguise what might appear to be the same characteristic. Therefore, to a Neolithic man, a fat woman was not ideal (statistically speaking). Also, he would be choosing an intelligent woman, which would only enhance the intelligence of his children (again, statistically speaking). Large breasts would show good health since, in poor health, a woman's breast size may decrease. A woman with a strong sexual desire for a man who had an athletic physique, high forehead, and non-protruding jaw (remember neoteny), but a well-shaped chin, would be selecting a partner with high intellect and sound physical fitness (statistically speaking). Therefore, a Cro-Magnon might have regarded a Neanderthal as "ill-bred.". It has been suggested that fascism is actually a manifestation of the "natural" (Natural dose not necessarily mean good) instincts of Germanic peoples (who come from Cro-Magnon man). Most philosophers have suggested that fascism or racism has to be "carefully taught" to the young before it appears, but the East German younger population have been very carefully taught to believe in racial equality. As far as being brainwashed by parents is concerned, it was the Soviet state that did the brainwashing. Joseph Stalin brainwashed children to distrust their parents, espousing the idea that the "good child" told on his or her parents if they said anything against the philosophy of the state. However, there is now a great upsurge in racism, not only in East Germany but also in Russia. By some "coincidence," it seems that Jewish people take the brunt of this racism. Is it such a coincidence that the Jews appear in their classical form to look slightly like the Neanderthals (of the Middle East)? If the reader remembers my comments on instinct, then he or she will know that I am saying that it does not take much to turn a person into a racist if the "right" stimulus is given.

One characteristic of the intellectual personality is a desire to think in straight lines, or linearly. Also, the intellectual prefers straight lines, squares, and triangles. Cro-Magnon's stone tips for spears were triangular in shape. Psychologists confirm that ego-dominants prefer these shapes.

I will make just a few more points about the Northern European. Northern Europeans are the only race of human who can properly digest cow's milk. (As mentioned prior, only one in ten Northern European adults has a problem with digesting cow's milk, whereas only one in twenty non–Northern European adults can digest cow's milk.) Also the European has lighter bones than people of other races. The bones are more honeycombed, which enables people to have the specific gravity required for floating on water. (This is a general statement, not a fact carved in stone.) This accounts for why Northern Europeans tend to monopolise the Olympic medals in swimming. However, there is a possibility that with the interbreeding of a South Saharan African and a Caucasian person, the resulting offspring is South Saharan African (or a hybrid South Saharan African–Caucasian individual who would be referred to as black). An Olympic swimmer with the athletic powers of the South Saharan African and the lightweight bones of the European could break many long-standing swimming records. (I am sure that this statement will make politically correct people happy. Bless them.)

On Arab and Jewish Races

One skull found in Israel was, according to anthropologists, a hybrid between a modern human being and a Neanderthal. When imitation flesh and skin were added to see what this individual looked like when alive, it became obvious that he resembled not so much a typical or Jew but a classical one. Although Arabs may look slightly different from a Jew, I believe they have a common origin I have already made my point about the Middle Eastern Neanderthal as an opportunist. Imagine for a moment a group of Neanderthals on the march looking for water, game, shelter, or whatever. They are all very tired and on their last legs. Some might say, "What's the use of going on? Let's face it: We are all going to die." "No," says the leader. "Have faith in God. He will save us; something will turn up." So, they keep on going, against all odds. Now, of course, there may have been many a tribe that kept on going but was unlucky and died out. Some were lucky and survived. But those tribes who had no faith and "gave up the ghost" died absolutely. The tribes that survived were the ones whose members had faith and kept on going. They where like Mr Micawber in that he said, "Something will turn up."

As I have already pointed out, Neanderthal man used seduction methods when hunting, much like a cat. Their lying in wait for prey involved the principle of the cat. They used the seductive self integrated with the hunting ego. Therefore, they were dominant in both self and ego, but, unlike in Peking man, the self and ego were integrated. (My theory only) Middle Eastern Neanderthal man may well have had an integrated BC profile, while modern Middle Eastern and Southern Asian people may have a BCD profile. I must again emphasise that this is only a relative generalisation when comparing one population with another. This Israeli Neanderthal also showed signs of religious beliefs, such as decorating graves with flowers and performing other rituals. One skeleton of a Neanderthal who lived to be an old age showed signs of early brain damage and physical degeneracy. It is unlikely that he was able to hunt; therefore, he was probably tolerated and looked after by the clan, much like a welfare state does for its weaker members. A welfare state maybe.

Religious people tend to exhibit wishful thinking or dreaming, as if wishing for something will make it materialise. Praying is a form of wishing without having to do anything constructive about one's situation. The idea is that God will do what one wants him to do. In fairy tales, there are many instances of impossible devises such as the crystal ball, the magic carpet, the magic mirror on the wall, etc. Today, we have these devices in the form of televisions, radios, cars, aeroplanes, etc., which are created by using science, or created by the ego. I shall delve into this farther on.

Starting from Scratch

Before I move on to the topic of civilisation, I would point out an important fact, one which may not be immediately apparent. Amongst different organisms, it is the lower organisms that are able to start life off from scratch. A lizard hatches from an egg. It is then on its own to survive the best it can. However, any mammal has to be looked after and taught how to survive before it can be sent out into the world to fend for itself. A monkey has to be looked after by its mother for many months and has to be taught not only how to get its food but also how to get on with the tribe. A chimpanzee is even more dependent on its mother – and for a longer period of time – than the monkey and probably has to learn even more than the monkey. A human is totally helpless at birth and depends on its parents until in its teens. And if a human is sent into the wild, it needs not only information but also weaponry, clothes, etc.

If a chimpanzee, and a human without any clothes, weaponry, or tools were forced into the jungle to look after themselves, then the chimpanzee would come off best. It would take some time for the human to make stone tools, spears, bows, and arrows, and also build a fire and a hut. In that time, the human being may be attacked by a wild animal – maybe even before he can even get started. However, as a matter of odds, if a chimpanzee and human were sent into the jungle, both with whatever they could carry from civilisation (gun, tent, medicine, tools, etc. The fact that the chimp would not know how to use these things is not the man's fault.), then the man would come off best. I believe this to be self-evident.

One cannot always judge an organism simply as it stands alone. An organism must be judged on its own terms, not by the terms set by another organisms. Consider the following analogy. Many a man has made the proud boast that he was able to build up a business from nothing, but that does not mean that he is better than other men if they all had a business to begin with. Many businessmen can manage to build up a business from scratch, only to flounder when reaching a certain level where they are at a loss about what to do next – or they may put their business in the hands of others who are more qualified in order to expand. But those more qualified to expand a business would not necessarily be able to build one up from scratch in the first place. It's all a matter of at which level one is best suited to live. The higher organisms are better suited to start off from a higher level. It is an obvious fact that a Machiavellian can start from scratch since it is better at seducing its way into a position to control the means of production, whereas a non-Machiavellian would not.

Chapter 9

Civilisation

Many an anthropologist has often wondered why civilisation occurred when it did. Why not sooner or later? It is possible that some small civilised settlements existed during the Ice Age, but these would have probably been situated near the sea. When the ice melted and the sea level rose, these settlements would have been flooded and the occupants would have moved on. Myths are often made from a variety of sources, converging over time to make the stories more interesting. It's possible that the Atlantis myth is a true story arising from old memories of a settlement that existed some time ago near the sea and was engulfed by the gradual rising of the tide. Storytellers often want to dramatise their stories to make their listeners more interested, so some may have merged the story of Atlantis with other stories of cities destroyed by volcanoes or the like. It has been suggested that the Egyptian Sphinx was built long before the pyramids. I do not consider this suggestion outrageous, as many traditional historians would, since humankind had the capacity for civilisation thousands of years before the end of last Ice Age. Even if the Sphinx was built during the last Ice Age or before, this does not mean that the people who built it were sophisticated and civilised or anything like the ancient Egyptians. They could well have been semi-barbaric. You don't have to be clever to be civilised any more than you have to be an idiot to be a barbarian. Civilisation is, in many cases, a matter of values. A barbarian may think it more important to be a warrior than a philosopher, where a civilised person would think the reverse.

The first signs of civilisation appear around Jericho about nine thousand years ago, at the end of the last Ice Age. Jericho's walls were built to protect the inhabitants. "From what?" one might ask. If a people were willing to invest so much labour into a single site, then it is quite obvious that they were there to stay and were not following game. It is therefore likely that they were farmers of sorts. I would suggest that the first civilised people were descendants of Neolithic and Palaeolithic hybrids. These hybrids may have wanted to own territory and therefore stayed in the same place – as, perhaps, Cro-Magnon (ego) – desiring to be a part of the earth (like the Neanderthal self) and study the plant life. Farming may have been initiated by a tribe concentrating on gathering vegetable matter during a time when prey was reduced in number. It would follow that it would be in these people's interest if more of the plants they gathered did not have to compete with "weeds" (other plants that they did not like). So, they may well have cleared the land of weeds and then allowed their chosen crops to multiply. It would only be a matter of time before they realised the association between seeds and new life. The groups that settled down to farm would have occupied a smaller area, less land per head of population (considering that they would defend their fields from unwanted herbivores or carnivores, the latter of which would prey on the few domesticated animals they may have had), which would have been easier to defend against other tribes. If there where a surplus of food, the community could have afforded to keep those men whose job it was to think and operate systems of counting and writing or to keep an army of specialist fighters. Of course, if they had surpluses, they could always trade with other tribes, especially if the settlement was on a navigable river, making it is easy to transport a great many goods to and from the home base. As a side note: It has been suggested that their taste for beer was the reason why human beings settled down to farm. Who knows?

Civilisation and Its Downfall

The trouble with civilisation is that it seems rather unstable. It appears that no particular nation stays on top for long. But I do not say that one should judge a civilisation (or a people) by how low it has (or they have) fallen, but by what it has (or they have) given to humankind. The Greeks gave humanity great ideas and a vision of greater values than ever before. The Romans continued to progress after the Greeks, not by destroying the Greek culture, but by respecting and leaning from it.

The British had an empire which took new ideas of personal freedom and liberty from around the globe, although these may not have been practised as perfectly as they should have been. But the ideas were still there. The Americans also created new ideas of democracy and technology, as well as Disneyland. Some historians claim that Attila the Hun and Genghis Khan and his "merry" Mongols were great men. But what did they give to the world except destruction and disaster? Their only testament was to show how nasty and destructive men can be, as they portrayed the lowest and basest in humankind. The formerly mentioned empires, although not perfect, showed how high humankind could rise. All people, if educated, are capable of living a civilised way of life and are also able to read and write, do mathematics, etc. I will suggest that some races unto themselves will evolve a form of civilisation different from other races' because they have a different personality profile type (taking a population as a whole, nothing is absolute.).

From the North

It would appear that all European civilisations were created by people from the north who intermixed with the local population. The Hittites came from the north (Hungary) and discovered iron. The Greek civilisation was started by northern barbarians, as was the Roman civilisation. It would appear that living in the north taxes people both physically and mentally, since northerners need to store and preserve food for the winter and their living quarters have to be strong to withstand the force of winter weather. If your house blows down or is destroyed for any reason, it is almost impossible to rebuild while in a sub-zero climate without shelter. Therefore, you have only one chance to build it properly during the summer months. Being taxed so hard in the north would mean that any northern people moving south would find life much easier, with crops growing faster and for longer periods of time. It's like this with the fictitious superhero Superman, who was born and evolved on a gravitationally strong planet called Krypton, which meant that his physiology genetically adapted to high gravitation. When Superman moved to any gravitationally weaker planet, he would have superior strength compared to the local inhabitants'. When the barbarians came down from the north, they, once used to the new food sources, found life much easier than the local inhabitants did. It is quite possible that in not having to compete with nature so intently, they may well have had plenty of energy to build for the pleasure of it, which opened them to the possibility of seeking challenge. Competing over who could create the largest building would certainly be a fast-track way to build a city.

Civilisation and Progress

Egypt

Many people may wonder why some civilisations progressed more than others. I believe that continuity of civilisation had something to do with it. Let's start with the Egyptians. Egypt was created near the Nile River, which meant, first, that its people had plenty of sun and water and a year-round growing season, which, in turn, meant that many people could concentrate into a small area (relatively speaking). Also, Egypt as a civilisation had length rather than width, which meant that those in power could easily control the population. On top of all this, the Egyptians enjoyed many thousands of years of an unbroken continuity in ideas, not losing their ideas or having them destroyed by invasion. They also had a population that could afford to support thinkers, writers, mathematicians, etc. I would also say that just because a people have been able to advance in this way does not necessarily prove high intellect on the whole. Although many Egyptians were highly skilled in various industries, this does not mean that such skills did not exist in Europe at the same time. In fact, if you were able to compare the standard of housing for the average Egyptian to that of the European, you would probably find that the European had far more artefacts and that his "huts" were more sophisticated than the average Egyptian's, for the simple reason that the European had to contend not with just the rain, but also with the cold snowbound winters. Also, the European's clothes would have to be more sophisticated, for obvious reasons. (I discussed this in an earlier part of the book.) Also, the large houses in which the wealthy Egyptians lived were made only by the cooperation of a large workforce, not because of higher intellect. The European families at the time had to build their own houses without any, or else with little, outside help. Let's suppose that many thousands of years in the future, some archaeologist digs up and compares today's Egypt with a country like today's Iceland, for instance. The archaeologist could say that Egypt as a nation was wealthier and more powerful and had a greater infrastructure than Iceland did. But Iceland is still part of the Western world (with a high standard of living, although its people live on a barren rock). Egypt is part of the Third World. Therefore, the reason why Europe was considered backward was only because of their lack of collective organisation, not because its people were short of innovative intellect.

China

The Chinese started their civilisation at about the same time as Westerners started theirs, but their civilisation was based on rice. The Far East Asian Race peoples are, by nature, herbivorous. They have longer stomachs than people of other races do and, therefore, can live quite contentedly on rice and other vegetable matter. This means that a plot of land in China can hold an even greater number of people. It was believed that the Chinese civilisation developed separately from the West, but it has been revealed that a tribe of Northern Europeans existed in a remote corner of China as many as four thousand years ago. This debunks the Chinese myth that the nation developed in isolation from the West. These Europeans took the wheel with them to China. The Chinese probably got the idea for the wheel from them. The Chinese, like the Egyptians, could support an upper class of clerics, et al. as well as a huge army. Any knowledge or idea that was found or that occurred in one part of China could, with time, pass to another area. It is true that China was invaded from time to time, but the invaders, like the Mongols, were small in number compared to the whole population of China. As China is a mountainous region, the invaders may have found moving large armies from place to place somewhat difficult. It would have taken them a long time to destroy all the knowledge possessed by the Chinese. Before they could execute that task, the Mongols saw the virtues of the Chinese civilisation. Instead of destroying what they did not understand, they simple took it over. This meant that all the ideas, which may have taken thousands of years to acquire, survived.

European Civilisation

It is accepted that Europe civilisation was born in Greece about 1400 BC. The first people who could claim any kind of civilisation in Greece were in Mesopotamia at about 3000 BC. These were the shore people, mostly fishermen and farmers. Nothing much happened in Greece until the European barbarians invaded from the north about 1600 BC. (I would point out here that the northern barbarians never became civilised by themselves. Only after they made contact with the Mediterranean people did they eventually became civilised.) The barbarians did not kill off the shore people but learned from them and made them part of their tribe. Only after that time, about 1400 BC, did the Achaean civilisation take off. This civilisation was to change the world.

In 1100 BC, barbarians from the north invaded again, but this time they where an Iron Age people called the Dorians. The barbarians would not leave the Achaeans alone, as they forever hounded and destroyed everything before them. The Achaean found that their bronze weapons were no match for weapons made from the new metal, iron, no matter how hard they fought. Eventually, in about 800 BC, the barbarians saw the virtues of Achaean civilisation. At that time, the Achaeans and barbarians settled down together to build up civilisation again. The result would be known as the Greek civilisation. The people had to relearn a great many skills and knowledge lost in times past. However, this was the golden age of the Greeks. They grew and prospered, not only in military power, but also in ideas about how human beings should live together. They invented democracy, expanded mathematics, and took on the Middle Eastern idea of phonetic writing and expanded upon it.

It was about 150 BC when the Romans invaded Greece, but they did not destroy knowledge or culture (although a city here and there may have been destroyed). It is quite possible that the Romans had gained the knowledge from the Greeks beforehand. The Greek–Roman civilisation progressed onward, becoming stronger with new technological ideas such as concrete, which the Romans exploited to the full. Unfortunately, the Romans did not know anything about gunpowder, which would have saved the West from a huge setback from its inevitable progress. The trouble with Europe is that it is more of a peninsular extension of Asia and less a separate continent. It is only referred to, and treated as, a separate continent for purely political reasons. Asia can support a huge population that can invade Europe. This is like pouring barbaric hordes into a funnel. The hordes grow stronger as they advance upon a smaller population of Europeans. It would be true to say that the Roman civilisation under Constantine moved eastwards, where most of the administration took place. This impoverished the western half of the empire under Rome, so the Roman Empire of the West was already weakened by the time the barbarians arrived. The final fall of the Roman Empire was due to the Hun invasion from the east. The Germanic barbarians pushed forward, with the obvious consequences. In 455 AD, the Vandals invaded Rome and may have destroyed the infrastructure and knowledge of that great civilisation. The whole of Western Europe fell to the barbarian before the barbarian could ever understand civilisation's virtues. This meant that many people were killed. Some starved to death since there was obviously no means of producing the same amount of food as organised farming could produce. Also, much of Rome's food came from North Africa, but there would be no way of paying for or organising distribution. The most important thing to remember is that the weaponry of the Romans was not that much more advanced than that of the barbarians. One thing that the barbarians had and that the Romans did not was the stirrup, which enabled the full use of the horse as a weapon of war. This invention led to the knights in armour of the Middle Ages.

So, after the invasion European civilisation had to start all over again, almost from scratch. Strange thing is that only twenty years after the sack of Rome, a barbarian king, Clovis of the Francs (a barbarian tribe), started the climb back to Western civilisation. It was only about the beginning of the sixteenth century when civilisation re-achieved the state of the Roman civilisation just before it fell. It is true to say that westerners did retrieve some of their former knowledge from the Muslim Arabs, who had gained that knowledge (to be fair to the Arabs, they added our present mathematical system, which they retrieved from India's Hindus) from the Eastern Empire when they conquered Constantinople previously. So, unlike the Chinese civilisation, the Europeans had to start over. The Chinese are free to boast about their great achievements, but they may forget how little time it took the Europeans to achieve great things, compared to the many thousands of years the Chinese had to do the same. The saving grace of Western civilisation was the gun, since only an organised civilisation would be sophisticated enough to make both gunpowder and the cannon. The more sophisticated the weaponry a civilisation possesses, the more sophisticated the conquering power needs to be to meet it. In any event, the point is that if one civilisation conquers another, this does not mean that the conquered party's knowledge and skills will be destroyed.

The Greeks were masters of mathematics but were encumbered by a difficult mathematical system. The Hindu Indians invented the mathematical system we enjoy today in the West. This system was introduced to the West by the Arabs. With the Greek mathematical formulas added to it, this system excelled totally.

There exists in this day and age a guilt complex about being a white European. Some Europeans feel a need to make a kind of apology for past successes. Pundits seem to grovel when faced with the idea of foreign superiority, making apologies for things that were boasted about in the past. It was believed in Victorian times that the Europeans were a superior race because of their technological ideas. They really went over the top in boasting of their absolute superiority

But although general truths may exist, no truth is ever absolute. One can only compare things in their context. If races such as the Far Eastern man achieved things before Europeans did, it was only because of the continuity of civilisation. In other words, the Far Eastern man did not see their civilisations destroyed and then have to build it up again virtually from scratch. Once westerners were able to secure their civilisation against the barbarians and became able to shrug off the dogma of an Eastern-based religion, their civilisation went from leaps to bounds unlike any other. The Chinese may boast that for thousands of years they have had the same or a superior level of technology as seventeenth-century Europeans. Why was it that Westerners excelled to their present level and on their own backs in just four hundred years but the Chinese did not?

It has often been wondered why there is so much tribalism in the world. "Wouldn't it be better if we were all part of one big brotherhood of man, where there was no tribalism and no nationalistic differences?" My answer is *no!* If humankind were made up of one unified mass, then we would have something like a state monopoly. Now, we all know what happens in a state monopoly. Nations tend to set up different systems of administration and government. However if nations do their own thing they can compare themselves with others, and see which systems works out the best. After World War 2, many nations wanted change. Free enterprise seemed like nothing more than organised greed and selfish competition. State control, even if it retained internal competition internally, did not take the place of free enterprise. I believe that for all its faults, nationalism is one way of uniting people who have a common identity. Just because there is competition against an outside competitor (whose people also have a common identity) does not mean that things must lead to war and destructive conflict. The opponents can, at best, compete in a non-belligerent, tongue-in-cheek way while retaining mutual respect. One should always remember that aggression is not the same as violence. Violence can exist without aggression just as aggression can exist without violence.

COMPARE IMPERIALISM AND EXPANSIONISM

The greatest amount of self righteousness comes from the American's about British Imperialism. The Americans did not practice Imperialism they practised "Expansionism" instead. What is Expansionism? Well that means you take over a Third world peoples land, exterminate the inhabitants to near extinction, which could be by encouraging disease, or exterminating their food source (i.e. killing off the American Bison) or generally "ethnic cleansing with extreme prejudice". You then encourage immigration to that land by your "own kind" which completely outnumbers the original inhabitants and then declare one man one vote. Then you proclaim this is a democratic "Sons of the brave in the Land of the free". It would be only fair to say that the British farmers' did practice "Expansionism" in Tasmania at about the same time, but we do not boast about it. I do not write the last paragraph in order to sit in judgement upon the Americans as a wicked people, only to make the point that their self righteous revelation about the British are totally out of order.

IMPERIALISM

That's were you take over a third world countries land, exploit the inhabitants by giving them jobs that they were only to happy to do, and make a good profit out of them, until the day comes that is, when the Imperialist realise the game is up, and they then give the colony it's independence and the people their land back. (With a ready made infrastructure.) Somehow I think that the North American Indians, the Indians in India, would choose Imperialism any day to Expansionism. Lets suppose the British having reached India had practised "depopulation with extreme prejudice" with the East Indians and then filled it up with British people. They could then declare one man one vote, they are "The sons of the brave in the Land of the free", and down with wicked Imperialism. Of course if the Southern States of America had practised "Expansionism" with the Negro in the last century they would be more idealistic in the eyes of the world during the fifties, and if White South Africa had practised "Expansionism" with theirs, we would think of them on an equally moral par with the USA. I would point out here that in no way do I sit in judgement upon the USA for past "sins" I am only making a comparative point in order to get the anti-British attitudes in America and around the world into perspective.

Chapter 10

Society: Its True Meaning

Mrs Thatcher once said, "There is no such thing as society, only individuals in it." This is the most foolish statement I have ever heard her make.

Society is one of the most important influences upon the individual. I remember when two child killers were analysed by a psychologist. The psychologist said that if either of the two boys were with another gang of children, then the killing would not have occurred. The boys created their own society of two. It is only that two similar minds got together and psyched each other up that made the killings possible. This is somewhat of a negative example. Most things that people have in common can be creative.

Every society has its own personality. A party of people who were older than fifty would have a different personality than a party of sixteen- to twenty-year-olds, just as a party of all women would be different from a party of all men. A party of all black people would have a different personality than a party of all white people.

People united by age, race, nationality, class, or gender all would create their own personality. Imagine a teenager being asked by his granny if he would like to attend an over-fifty party. What would his reply be? "O! No, I don't want to mix with a bunch of oldies." Does this mean that he hates all old people, even though he may love his grandmother? Of cause not. It is simply that he is objecting to a party with an over-fifty personality. Not every attendee would have a bad personality, in his view; it's only that he does not want to attend a party with an over-fifty personality. So it is with a white person not wishing to attend a black party, even though the white person in question may have a black friend. One black friend is not a society.

Religion:

I use the term *religion* in the context to mean "organised belief," all people believing in the same thing because that is what they were brought up to believe, without giving it much thought or using reason. A person can believe in a god without either having collective belief or belonging to a church. Also, a person can be an atheist and still be brainwashed in ideology (think of communism).

Why do we have religion? Well it starts off with a need to explain things about the past, like where do we come from? etc. how does the world operate the way it does? Science has for the Western world answered most of these problems, at least enough for us to realise that you do not need a God to explain all our questions.

However there is another reason for religion. I remember the case of a Mohammedan women saying that she needed the Muslim religion because it gave her such solemnity. Why would she need such solemnity, I am not suggesting that you can judge a people by one woman's statement, but I do believe that solemnity has a great deal to do with the Muslim religion. The term Muslim means I believe, "submission to Allah".

Civilization means living with strangers, although this may be easy for northern light skinned people.

As I have mentioned before mankind may not like living amongst dark skinned strangers for reasons I have already defined.

Living with dark skinned strangers may cause a perpetual state of tension, therefore stress in the individual.

If a person in the West suffers from stress, for whatever reason, the Western doctor will prescribe drugs of some sort to relieve the symptoms of stress.

Is it such a coincidence that a communist in the past described religion as the opium of the masses.

Is it possible that the solemnity of religion is necessary for dark skinned people to live amongst strangers.

I do not carve this statement in stone, but I would like the powers that be to question the possibility of this.

I have heard an African say "We prefer to live in small communities rather than cities". This is only a comment from one African, (this is anecdotal which I do not offer as proof). Although Africans may live in cities it does not mean that they are happy doing so.

Why do we hear of so many people in Africa living in areas of where there is very little clean water to drink? Why don't they move towards the cities where there would be clean water and if starving they could be more easily fed by world aid?.

May be they prefer to risk starving than having to live amongst dark skinned strangers? I really don't know.

Prejudice

Prejudice is defined as "a *conclusive* opinion formed beforehand or without due examination." I use the term "conclusive opinion" in contrast to a strong possibility. An individual may be aware that what he or she thinks may be wrong, but he or she may simply be playing the odds for the moment, lacking the information necessary to form a different opinion (the establishment may withhold some information from an individual). When many ethnic or non-ethnic people claim racial equality or racial inequality, is this an opinion formed through long study and thought about anthropology, or is it simply what the person wishes to be true?

Bigotry

Bigotry is defined as "being blindly and obstinately devoted to a particular creed, idea, or party, irrespective of any opposing facts or evidence, circumstantial or otherwise." *Sounds like religion to me.*

To help you understand bigotry, I would like to tell you a hypothetical story. Let us say that there existed, many thousands of years ago, a tribe of people who followed a prophet. The prophet said that God had told him that the world was round like a ball. God also said that the sun did not travel round the earth but the earth around the sun, and the earth spun around on its axes, which caused night and day. Some asked the prophet, "How could this be so? If the world were round, then the sea would all run over the edge. Not only that, but when the earth spins around away from the sun, why do we not fall off?". The prophet said, "You are not to question the great God. Anybody doing so is guilty of heresy and therefore will punished for bringing bad luck upon the tribe." The tribe, not wishing to upset the prophet or the great God, especially since the prophet had a bunch of strong men to back him up, agreed to teach their children whatever the prophet said. This became the tribal religion.

On the other side of the mountain, another tribe heard about the prophet and his revelation but refused to believe him, sticking to what they considered common sense. The first tribe was right for the wrong reasons, but the second tribe was wrong for the right reasons. The first tribe was a tribe of bigots, and the second tribe was a tribe of reasonable people.

The point I wish to make is that people have a right to be wrong without being accused of bigotry when their reasons are what they consider to be common sense and do not contradict what is known as scientific fact at the time. Let's suppose that it was found that God actually made the universe in seven days and that fossils in the earth were tricks of the Devil. Would this make Darwin and every evolutionist a bigot, and every religious fundamentalist a person of reason? Of course not, because the Darwinian evolutionist believed because he or she had a reason, not a dogma, to do so.

Bigots will rarely argue through a point. If the point of argument appears to go against them in debate, they tend to either ignore the issue or change the subject. This is to give the impression that they are above arguing and that their opponent is not worth arguing with. The real reason for this behaviour, however, is that the argument is going against them and they don't like it. One strange thing about the fictional "Alf Garnet" character on TV was that he was at least willing to argue things out, even though his character was somewhat ill-informed. I would say, however, that the Alf Garnets of this world are far easier to reason with than either politically correct people or any religious bigots. Alf Garnet's greatest virtue was that although he did not have the greatest mind on earth, he at least had a mind of his own, which is more than can be said for people who subscribe to PC or any other belief system.

Where physical appearance is concerned, suppose, as a hypothetical point, we had visitors from another planet who looked like insects but who were, by human standards, perfectly civilised, kind, compassionate, and considerate. Would it be bigoted if many people said that they would not work with these aliens because they could not stand the sight of them or to be in close proximity with them? I think not. The fact is people cannot help what they honestly feel. Bigotry is not about how one feels or what one thinks. Bigotry consists of confusing what one thinks with what one feels, and vice versa.

Point about Proof

It could be said that there is no proof that UFOs do not exist. Well, fair enough, but there is no proof that they do exist, either. So, do we therefore assume that UFOs are just as likely to exist as not? No. In such matters, one goes by what one knows, according to the general circumstances of one's day-to-day life in the world around us. This point is accepted by the scientific community as a whole. It does not mean that if UFOs were to turn up someday, the whole of the scientific community would be proved to be a bunch of moronic bigots.

Back to the subject of religion, I do not say, to start with, that people who believe in God are bigots. As an agnostic, I do not know whether God exists or not, and I do not know of any scientific means to prove or disprove the existence of God. So, when I put down religion, I am not criticising belief or having faith in God. I am only criticising mindless human-made rules (supposedly God's) that are imposed upon individuals or societies without giving people the right to question those rules, which may have no relevance in this day and age. It is normally the religious point of view that anyone who does not believe in God is in denial of and against God, and that his or her soul is in danger. I would say to those people that just because I do not know whether UFOs exist or not, it does not mean that I am against UFOs, that I have got it in for UFOs, or that I hate UFOs. It simply means that I do not know whether they exist or not. Savvy?

Religion

I remember reading an article on the Internet that said that Muslims are not happy living in Egypt, Syria, Iraq, Iran, or any Muslim state. They are happy living in Australia, Canada, the United Kingdom, Germany, or any non-Muslim state. After immigrating to a non-Muslim state, some of them then want to turn the non-Muslim state into a Muslim state. Perhaps they want to be unhappy living in a Muslim state? The reality is they cannot understand that it is the power of religion that makes an unhappy state.

Religion is supposed to be about belief, not about scientific fact, but, unfortunately, this is not always the case. Let's suppose a man in the centre of a crowd of people were to say that he could defy Newton's law of gravity. He holds a hammer in his hand and says, "I will make this hammer fall upwards away from the earth." The crowd laughs at the "joke." The man releases the hammer, which floats upwards to the sky. The crowd, aghast, probably claps at seeing such a fine "trick." The man at the centre says, "It's no trick. I simply defied gravity."

Now, the crowd believes that he is insulting their intelligence. "We all know that gravity cannot be defied in this way." The crowd insists that it must obviously be a trick and becomes angry with him for trying to "brainwash" them into a contradiction of something which they know cannot be contradicted. Unfortunately, religion and personal beliefs that remain in a culture without argument or contradiction can have a similar effect. If a closed culture has strong views on an issue and everyone agrees that they are correct without considering any alternative viewpoint, the people will see their views almost like a scientific fact. Anyone contradicting them will be seen as a trickster, a conman trying to brainwash them into accepting a contradiction of "common sense," no matter how convincing his argument turns out to be. This I have found to be the case when debating with religious people or people with strong political views.

The Criminality of Organised Religion

I will always remember the time when, in a film of the Beatles, Ringo cracked a corny joke. John Lennon exclaimed, "Ringo has cracked a joke, so let's have some organised laughter."

So, John and the other two Beatles cried quite mechanically, "Ho ho ho ho ho," in response. Now, of course, the very idea of organised laughter is a joke in itself, since it is a contradiction in terms. Laughter is something purely spontaneous. You cannot force someone to laugh. A person must think sincerely think something is funny before he or she will laugh at it. Any other laughter is only acted out and, therefore, false and artificial. What about organised love or organised hate? Aren't these things equally contradictions in themselves? How about organised thinking? Now, I know that one can organise his or her own thinking, but to attend a class and witness organised thinking in the sense that the object is for everyone to come out thinking in the same way is nothing but mindless brainwashing (the correct term would be *brain conditioning*). Belief is a combination of thought and feeling. Therefore, organised belief is as stupid, mindless, and downright ridiculous as organised laughter. The most significant thing about religion is that not only is the individual lost to reason, but he or she then imposes that very religion onto his or her children, expecting unquestioned obedience to the dogma.

These children grow up with just as much determination to impose that religion upon their own young, demanding equal unquestioned obedience. It is almost as though religion is a virus continually spreading its structure like a disease. I have heard the saying "Give me a child until the age of seven, and I will have that child for life." This statement fills me with horror, knowing that a child can be conditioned, like a machine is programmed, at such an early age. I honestly believe that any religious instruction of children below a certain age should be made a criminal offence, since religion opens a child's mind to abuse. It appears that the religious lunatic fringe is only outraged if a child is "carefully taught" racism at an early age.

The difference between a human being and an animal is that the human has the ability to think consciously about new conditions that appear. If a person is raised from childhood to believe in absolute dogma and to show absolute, unquestioned obedience to that dogma, then how different is he from an animal which behaves with absolute obedience to its instincts, which it inherited genetically from its parents, and which, in turn, it will pass onto its young?

I believe that religion is a form of culturally inherited instructions which take the place of instincts. It serves as a means to regress to our animal selves by supplanting cultural dogma for instinct, which, I believe, is somewhat dangerous in the long term.

Also, it is believed that a religion of some sort is necessary to introduce order and discipline into civilisation. Intellect might be the medium by which some people choose to discipline themselves. But, of course, not all people or societies are lucky enough to be gifted with such a precious commodity. I mean, it is only a matter of common sense that disorder and mayhem are not in the perpetrator's interest in the long term, since it is only a matter of time until the perpetrator becomes the victim.

There is the old saying: You can't put an old head on young shoulders. Well, I believe that religion is a crude way of trying to do this. By creating dogmas, the older generations are trying to impose the benefit of past experiences on the young, who do not simply take the older generations' word for it. If the older people tried to explain things pragmatically, then it might become clear that they do not understand why things work the way they do when people follow such rules, as the older people themselves inherited those rules from their forebears.

Perhaps it would be a good thing if the rules of life were defined logically. Then, we would not need religion to impose rules. But then that would mean that people would have to think for themselves, wouldn't it? The trouble with religion is that if it is necessary for a people to have a religion to keep themselves in order, then that religion can be an encumbrance when having to deal with social and technological changes. If a society did not need a religion for social order, then its people could concentrate on creating programmatic solutions to new problems. The Greek philosopher Epictetus once said something like, "People don't think; they merely follow trends."

Every new religion or philosophy is based on a desire for freedom and the destruction of tyranny. What happens, however, is that in people's hating what they are against, they tend to turn the new idea into a new tyranny. After the fall of Rome, the barbarians created havoc with any emerging civilisation, what with attacking towns and villages and making blood sacrifices, etc. It is no wonder that the European people were willing to ally themselves to the Christian Church for purposes of their own defence. The attitude developed that Christians were good and pagans (non-Christians) were bad. This worked fine while it was necessary. As time went, pagans no longer attacked Christians, but Christians' hatred of non-Christians still burned in the folk memory, which reinforced the idea that it was necessary for Christians to unite in defending Europe against the Islamic Empire. The Pope, as the head of the Christian Church at the time, was in an ideal position of power. The whole of fearful Europe was only too glad to accommodate him. However, power tends to corrupt, and absolute power corrupts absolutely. The Pope and his cardinals, bishops, et al. fell into disrepute amongst those who kept an open mind and their eyes wide open. When Protestantism raised its "ugly" head, there was bound to be backlash from the Pope and his cohorts. They regarded Protestantism as a rise of paganism and a form of dissent. The Protestants regarded the Pope, bishops, cardinals, et al. as the oppressive Devil who would not allow people the right to think for themselves. Now, the Protestants (especially the Puritans who fled to America to escape tyranny), believing themselves to be under attack and distrusting anyone who was not a Protestant, became tyrannical (performing witch hunts, e.g.) just like the Pope and other high-up members of the Catholic clergy. I conclude from this that each fight against tyranny becomes its own tyranny. In Northern Ireland, there is the "tyranny" of the Protestant majority over the "oppressed" Catholics, whom the Protestants see as a tyrannical Catholic minority trying to oppress them by forcing them into a tyrannical fundamentalist Catholic state of Southern Ireland. The same thing can happen with capitalism versus communism.

The Islamic Empire saved many people from slavery in the beginning and was therefore seen as a liberating force, the light of the world. As a result, Muslims became filled with total confidence that Allah was on their side, which meant that they could justify almost anything, including using tyranny in order to wipe out tyranny. In France during the 18th century the French people where treated badly by the aristocracy which lead to the French revolution. The same thing happened with communism. The reason why communism emerged was to liberate humankind from the tyranny of exploitation. But in doing so, communism created its own tyranny. I would say that a certain amount of tyranny to oppose tyranny may be excusable and justifiable to start with, but, like with all things, one must know when to stop. For example, when the Second World War started in England, the government imposed measures that people would not have tolerated in peacetime. These measures were lifted when the war was over.

Narrow-Mindedness

Narrow-mindedness is a term people like to use to describe any kind of moral restriction that occurs in society. I remember when society became more permissive in the mid sixties. Anyone against this permissiveness was called narrow-minded, as if being broad-minded meant you followed a particular philosophy. If you followed another, then you were narrow-minded. Well, that sounds, to me, a little narrow. I would say that a better definition of broad-mindedness would include the idea that although one may follow a particular philosophy, it is not written in stone and can be varied under different conditions. Also, it could mean that one has two contradictory opposing points of view which one believes are both equally true, especially when it comes to science versus morality.

Here, I provide further explanation. It is a known fact that human beings are gradually genetically degenerating. (While living a healthier lifestyle would make people healthier and also make them look better, this would merely compensate for or cover up a genetic weaknesses.) While there are many disorders involving a great deal of treatment which may be costly to society, this does not stop people from breeding and passing these disorders to future generations. This means that the human race will not evolve any further or become any better than it is today unless those who have fitter and healthier genes have more children than those whose genes are unfit and unhealthy. At the same time, it could be argued that it is difficult to stop people from having children or that society should let people die of starvation when they became unable to support themselves and their children. This is a dilemma which future generations will have to resolve. More on this later. But you can see how matter-of-fact problems come up against day-to-day morality, especially when only recently a "nutter" in Germany tried (rather incompetently) to practise selective breeding. Now, any suggestion of selective breeding throws otherwise normal people into fits of hysteria.

I return to the topic of the permissive European society of the 1960s. I remember as a fact that many a young woman lost her virginity after succumbing to moral blackmail "What's the matter with you? Are you a narrow-minded prig?" a suitor might have asked her, as though only sexual permissiveness were broad-minded and as though holding any other point of view was somehow being narrow-minded. The general defence of so-called broad-minded people is to suggest that obeying certain rules is wrong if there are circumstances where disobeying those rules will not necessarily cause any immediate detriment. Example: One might ask, "Why stop a person from taking drugs since many people take drugs and experience no harmful side effects? Plus, not all drug users end up as drug addicts." It could be said that it is narrow-minded to assume that taking drugs is bad for a person. What politicians normally say is that the mass taking of drugs would mean, statistically speaking, that more drug addicts would exist, which would not only cost the National Health Service more money but would also cause a rise in the crime rate and cost a lot more in resources which the country can ill-afford. Just because some people can get away with drug taking does not alter the detrimental effect that widespread drug use can have on society. Also, many people can drink and drive for years without causing an accident. Does this mean that drunk driving is OK, then? It could also be said that, to a lesser degree, permissiveness had a detrimental effect on society. I would not take away people's personal freedom because of this, since they only hurt themselves. Still, one should always keep an open mind to opposing viewpoints.

When early human beings found alcohol, they probably experienced the same effects that the American Indians, who were not used to alcohol, experienced. But over a period of many generations, those people who could hold their drink the best survived to have more children. Maybe drugs, in future, would undergo the same evolutionary process if it were not for the welfare state. It's all a matter of how you look at it.

I would like to make one more point. It is said that one should be open-minded. Well, an open mind can end up full of garbage, just as a closed mind can end up empty and useless. There is a third option: a discriminating mind which absorbs knowledge. The intellect discriminates between wisdom and garbage by employing thoughtful reasoning, which, hopefully, keeps the garbage out.

Chapter 11

Racism

Defining the Different Types of Racism and Intolerance

Racism is never a simple issue, although many people would like to have a short and sweet answer to it. Like being frightened by one's golliwog when one was young, or perhaps bad potty training, or some subtle action such as noticing their parents tend not to mix or talk to coloured people etc. I have heard that sexual jealousy is the reason of racism. I have also heard that if there were not any blacks, then someone would have to invent them. Maybe blacks were invented so that white people would have someone to be sexually jealous of? OK I am being sarcastic.

It is generally known that those people who are accused of racism are often denigrated by the media and the establishment, as though this will solve the racist problem. Let's suppose that one wishes to deter people from having heart attacks. We could say that the cause of heart attacks is being overweight, smoking, eating the wrong foods or too much food, not getting any exercise, having stress (which is caused by fear), etc. Therefore, one could say that people who suffer heart attacks are greedy, lazy, self-indulgent, cowardly pigs who lack moral fibre.

You see how easy it is to denigrate people if it suits one's political aims. If you wished to denigrate people who expressed racist feelings, it should not be too difficult. The greatest red herring about race is the issue of skin colour. No one, not even the most extreme racist, actually believes that the enzymes produced by the genes that lay down pigment in the skin have any effect on an individual's intelligence or personality. It is quite obvious that albino South Saharan Africans are hardly more likely to be higher in intelligence or different in personality than those South Saharan Africans who are not albinos.

Anthropologists have used skin colour for labelling purposes, not to classify gene pools from different geographical areas. When I use the term *white,* I mean European Caucasians; *black,* South Saharan Africans; *yellow,* Chinese and Japanese people; and *coloured* or *ethnic,* non-European Caucasians. The term *mixed Northern European* refers to predominately original Neolithic man or Germanic man with a small percentage of Mediterranean man, who, in turn, is a Neolithic man with a small percentage of Middle Eastern Neanderthal genes in his make-up. That's to put it *simply.*

I wish to point out that the small percentage of Neanderthal man gives flair, imagination, and wish fulfilment to the human being, as I mentioned earlier. I would point out that there is no such thing as an "English race" since the English are made up of Indo-European Celts; Romans, who were themselves from a mix of Germanic and Middle Eastern man; Angles and Saxons, who were people of Germanic tribes; and Normans, who were a mix of Germanic and Gaelic (Celtic). That is, of course, to put it *simply.* As far as racial purity is concerned, I only use the term in the relative sense.

Suppose a man claims to have a gold watch. He may be corrected if his watch is gold-plated, but gold-plated watches are often called gold. If a man says he has a solid gold watch, he can be corrected if, in fact, a certain amount of silver was mixed with the gold. What he means is that his watch is predominantly gold and not gold plate. Now, of course, a man could claim that a bar of gold bullion is pure gold – as near to pure as it could be reasonably made, perhaps. If you wanted to be technical, you could say that even gold bullion may contain a few impurities since it is quite impossible to get rid of every last atom of impurity, even in gold. Does this mean that anyone who claims to have pure gold is a bigot? No, you have to take what people say in context and stop trying to catch people out by nitpicking. So, when I use the term *racial purity*, I do not need to be corrected by a politically correct bigot.

The first point I will make about racism is that there is a distinction, namely how immigrations affected the host country. These immigrations should be treated on a political basis. Racism comes into play when we talk about how we treat people on a personal, one-to-one basis.

How we treat people on a one-to-one basis and how we should try to understand the feelings of individuals who are part of minority groups are perfectly legitimate points of political correctness. I would agree that making monkey calls at South Saharan African players from the football terraces is totally unnecessary and achieves nothing but counter-racial abuse. Neither does racial violence solve any problems. It only inflames a situation which could otherwise be resolved by using reason.

Would you say that monosexuals such as heterosexuals and homosexuals are narrow-minded sexist bigots and that bisexual are open-minded, liberal, intelligent beings? Of course not. Feelings are not evidence of bigotry. Only when one confuses feelings with fact does bigotry occur. Just as one can have sexual preferences, one can have sociological preferences, i.e. one may choose the sort of society in which one wishes to live.

First, let's suppose that a politically correct person were to say, "I hate crowds." Does this mean that this person hates the individuals in the crowd? Of course not. He or she simply does not like being amongst a lot of strangers. Now, one could say that a crowd is made up of many individuals. Is this a contradiction? Well, obviously not. But this is the sort of argument that politically correct people offer. Each person is willing to tolerate a different number of other people in a crowd. Now, suppose that such a person who had the same feelings about all races travelled the world. Would this make him or her a bigot? Well, no, of course not. However, suppose that this person found him- or herself amongst strangers of a new and unknown race, and was surprised to discover that he or she did not mind being in this crowd of strangers, without knowing why. Does this mean that such an individual has become a bigot? I will let the reader decide that for him- or herself.

Racism is not about hatred of people of different races any more than male homosexuality is about hatred of women or heterosexuality is about hatred of one's own gender. Just as there are sexual preferences, so there are social preferences. One may choose what sort of society in which one wishes to live, which may include the race(s) of people with whom one would prefer to coexist.

The racism that I will discuss is about what is simply felt within an individual in a host country and the effect that race can have on a nation and its culture. I have no wish at all to hurt a person of a foreign race, even if such a thing is the unfortunate outcome of my discussion. It has come to my notice that people call racism a disease of the mind or a cancer. Well, I consider this to be a gross distortion of events and language. If racism does spread like a disease, then it's because the discomfort of race is already there to begin with. But people *may* tolerate many an irritation in order to live a quiet life. (*One should always be aware that what appears to be tolerance might just be apathy.*) If they know of a movement that can solve the problem of racism, then they might be tempted to jump on the bandwagon.

Exclusive, Inclusive and Transclusive

When people use the term exclusiveness they of course only some people allowed in, but others left out. No problem in understanding this.

Now inclusiveness has been used in the past to describe a situation where people are not allowed out without permission. However inclusiveness has been used by the PC and the left wing of politics to describe to be able to come and go.

Know it is strange that the PC and the left wing do not seem to have a word for not being let out like the Soviet Union or any communist state.

However I would use the word Exclusive to not being let in.
I would use the word Inclusive to mean not being let out.
I would use the word Transclusive to mean, to be able to come and go.

South African Aper tide was exclusive, that is to say although they had their Townships, they were not allowed in white only places. However it was not inclusive like the USSR, they were free to leave the country at any time they wanted to. I would also point out that Spain from 1939 – 1975 was rather a nasty dictatorship under Franco, however no innocent person was prevented from emigrating if they wished to, unlike the inclusiveness of the USSR. Therefore I consider right wing dictatorships have the moral high ground compared to a Communist state. I think I would rather be a black man in South Africa during an Aper tide, then a North Korean at the moment.

Types of Racism

People are capable of racial feelings, racial prejudice, and racial elitism. They may see people of other colours as a bit too different. Also, the tribal amity and enmity factor comes into play, but there is a difference between these forms of racism. Only by analysing each in turn, in isolation, can we come to terms with the issue.

Racial Prejudice

Racial prejudice is to do with ideas that have been consciously learnt (or carefully taught). These ideas may be generalised truths or else untruths and distortions which have been picked up on one's journey through life. I do not believe that this sort of prejudice lasts long in the face of intelligent people, so long as the countering intelligence does not sound like self-righteous political correctness. Still, if a person has been "got at" by the age of seven, I have heard, then these prejudices may be extremely difficult to eradicate, as most people on the "religious kick" are aware.

It is known that many male birds of paradise are chosen by females for their great, spectacular plumage, which may inhibit them if confronted by danger. However, there is an evolutionary dead end that says that if a female in a safe environment prefers uninhibited fitness over appearance, then she chooses a male who will give her offspring fitness characteristics but not necessarily good plumage. Other females would not choose to mate with this male. Therefore, if a female chooses uninhibited fitness over plumage, she may produce sons that may be rejected by other females, then it would not be to her advantage in a safe paradise. There is a similar outcome when a person mates with a member of an out-group. Any children produced from this union would be rejected by their own group, no matter how fit the parent, a member of the out-group, may be. There would be social repercussions for one who associated with an out-group that he or she *believed* people of his or her own group did not like. Because of this, one may choose not to associate with the out-group for fear of being ostracised from his or her own group.

I would point out here that there is a difference between incidental prejudice and prejudice that is consciously taught by using the "drip, drip" effect. When youth in the United States are taught about the history of the American fight for independence, the British are obviously presented as the "bad guys." However, the average United States citizen is not brainwashed into thinking that the British remained bad guys forever. Also, when British children read *Noddy's Toyland Adventures* and see the Gollywogs as the bad guys, their minds are not irreparably corrupted with the belief that black people are bad.

Enmity and Amity

Robert Ardrey made a strong point about the phenomenon of enmity and amity between groups in the wild, which I mentioned earlier when discussing primitive humankind. Politically correct people make a great thing about this concept in order to trivialise the racial feelings involved. If members of a tribe do not like one another, if a state of disorder arises as a result, and if the tribe is then threatened by a foreign people, then there may well be amity within the tribe because of the enmity from outside. This is bound to inspire tribalism, of sorts. But in many cases, a people may be living together without any social problems to speak of. Problems occur when a foreign race or foreign group is introduced. Such is not a case of enmity or amity. This is a case of having genuine feelings of racial intrusion. I believe that the reason why English people are considered civilised but also so reserved is not entirely an accident.

When one is reserved, it does not mean that one is hostile or unfriendly. It simple means that one keeps to oneself. In being reserved, there is a lower chance of hostility and conflict arising from the misunderstandings that can so easily occur. I suppose we all know by experience that the best neighbours are those who keep themselves to themselves (one sees them only now and then) but upon whom one knows that he or she may rely on in an emergency. If an introduced group has a "Love thy neighbour" philosophy and perpetually becomes too familiar and friendly, then it could upset the status quo, thereby unwittingly causing disruption in that society.

The English Home Counties have perfected a reserved culture, at which many "intellectuals" scoff.

Ignorance

It is often said that ignorance or lack of intelligence is the cause of racism. I do remember an advert which said that one does not find racism in a maternity ward. Well, as it happens, a maternity ward has patients, namely infants, who have the least developed intelligence and the greatest amount of ignorance. The only thing that comes out of a maternity ward is a lot of noise and a lot of baby shit (at times a classical example of political correctness?)

Of course, the PC person will say that racism is caused by half knowledge. Does this mean that PC people know absolutely everything or absolutely nothing? Speaking of ignorance, the PC people I have heard on the TV hardly know anything about race, human evolution, or genetics, unless it is selectively in their interest to know certain things about these topics.

It is pointed out that ignorant people are more likely to be racially abusive than educated and respectable middle-class people. Seeing how male building workers are more likely to make abusive sexual remarks to passing women, does this mean that building workers have greater libidos than bowler-hatted office workers? I hardly think this is the reason that this type of abusive behaviour comes from the former group and not the latter. If ignorant people are more racially abusive, then it's because they are more abusive about everything, not just race. It is not a reflection of one's feelings, only of how one expresses them. Need I say more?

Elitism

Racial elitism entails the suspicion of others' racial inferiority, where characteristics about a people give a person the impression that those races may not be as evolved, in Neolithic terms, as one's own – and, therefore, the racial elitist believes it possible that interbreeding with people of the race in question may not be beneficial to his or her own race or genetic code (this is very politically incorrect).

Some characteristics of many races lead some Europeans to suspect that people of that race are primitive. I have already given many facts about neoteny that would make clear the reason why some Europeans believe that foreign races are primitive. Prognathism, a protruding of the lower face, the relatively large upper and lower jaws, is one such non neotenous characteristic. Also, the facial bones of someone with prognathism appear rather thick and crude. This lends an individual an apish, or obviously Neanderthal, appearance. This and the fact that South Saharan Africans in Africa and Aborigines in Australia are found in a primitive state buttress the position of the racial elitist. Although many South Saharan Africans may become doctors, lawyers, scientists, and so forth through very hard work, it is rare for them to excel in these professions (I am not suggesting that such a thing would be impossible). Neither are there any great revolutionary inventions amongst any ethnic group, in spite of the proportionately successful Asians. (Again, I do not say that such a thing is impossible. I hope that the politically correct do not find the last sentence too difficult to understand.) It does appear that only a small minority of South Saharan Africans are capable of inspirational invention. I have however seen some very clever examples of inventiveness by South Saharan Africans on TV documentaries. Nothing is ever absolute.

It could be argued that Caucasians have more hair than people of other races do. Does this mean that white people are more primitive? It so happens that anthropologists do not determine primitiveness by the amount of hair any more than by skin colour, since hair and skin colour are superficial characteristics. White women have no more hair than women of ethnic races. Hair on male Caucasian is a sexual characteristic that overrides the effect of neoteny, as I mentioned earlier. If protruding jaws were an environmental adaptation that overrode neoteny, then the South Saharan African's and Aborigine's bone structures would be more refined in their make-up. Also, the overall skeleton of the South Saharan African is more robust and, *in general*, slightly cruder than the European's.

South Saharan African Women do not seem to be any different from male South Saharan Africans in this respect, other than the normal relative gender difference. This may be the reason why it is rather hard for white people to accept racial equality except to acknowledge that is against the law to discriminate on the basis of race and except when blackmailed by a pressuring media. If anyone said anything that was not politically correct in front of the media, then that person would be morally crucified.

I have often thought that doctors and lawyers nowadays do not inspire the same awe and respect as they used to. Is this because coloured people who became doctors and lawyers did not raise the status of coloured people, but simply lowered the status of doctors and lawyers? OK, I know this does not sound very nice, but I believe this should be brought to people's attention. The more we acknowledge our deep-seated attitudes, the better we can analyse whether there is truth or distortion.

Colour

One result of people's racial feelings is that they are intolerant of people of other colours. I already mentioned the case of black male gibbons and blond female and young white gibbons. The point about the male gibbons being intolerant of other males is that they seeing blackness as something intolerable belonging to a creature that should be "kicked out" of their territory as soon as possible. Those gibbons that had the most intolerance for "blacks" stood a reduced chance of his mate's being "messed with" by any other male, which meant that his own genes would pass on to future generations. I am not suggesting that the gibbon is sexually jealous of its neighbour or that it knows or believes that there is any threat to its genes. It is simply responding to a desire to eject a black being out of its territory. The jealousy involved is more likely territorial jealousy. I have already suggested that our own ancestors may have possessed the same assorted colouring as the gibbons' – and for the same reason. It is still possible that, even after millions of years, people retain an intolerance of skin or fur darkness. Maybe we are stuck with it for good.

Simply Being Different

When a person of one race commits an act of violence against an individual of a different race, I would put this down to bullying rather than to racism. Still, the fact remains that a person who is different can attract the attention of a bully, especially if that person thinks that no one will give him or her moral support. It is very possible to have racist feelings, that is to say, for white people to want coloured people out of their territory, without any initial desire to physically or mentally hurt the people whose skin colour is dark. But, as is the case sometimes with frustration, a person who initially does not want to hurt the minority group may eventually join the bullies, since this may be the only way of expressing their pent-up frustration.

I believe that this is how fascism grows in strength. I would also point out here that there is a difference in belonging to a minority group which one identifies with and, on the other hand, simply being a different individual who rejects any allegiance to any group at all. When a person is an individual belonging to no particular group, he is not considered a threat to the group, but he may not arouse sympathy if bullied. If a person identifies with an outside group, then he or she may be thought of as an enemy within. That is to say, if the other group were hostile, then the individual may be under pressure from his or her group to work against the home group. Therefore, this person is a threat. This sort of thing happened in England centuries ago between Catholics and Protestants. You can now see how complicated human group (tribe) conditions can be.

Poverty

It has been pointed out by a newspaper columnist that racism is caused by poverty, given that there is no racial strife between Jews and Arabs in Kensington, for instance. Fair point. But when one is rich, one has more personal space, greater territory, and personal privacy. One has far more control over one's environment as well as better protection from intrusion. Also, Jews and Arabs in Kensington hold no power over one another, and both have the protection of the society in which they live. If one is very poor, then one may live in a one-room flat and have to share facilities, such as a toilet, with others. This is bound to cause frustration and stress. If one lives in a terrace house or within a block of flats, then one must contend with the noise from the neighbours or their complaints. Or, one's door may open directly onto the street or the flat's walkway, which means that anyone can stand outside the window or door and make a sudden attack without having to cross any personal territory. Having wealth means that one can live in a wealthy area with other wealthy people who generally have no temptation to steal from one another. One may also live in a cul-de-sac where there is no through road. Therefore, anyone wandering in that area needs an excuse to be there. In poor white areas, it is known that life can be tough and intolerable enough as it is, but most people who live there have social cohesion and cultural continuity. But when people of a foreign race and culture are introduced into the just-about-tolerable conditions, there is bound to be trouble.

With all the trouble with high-rise flats, it could be said that the architects didn't know what they were doing in the sociological sense in that they were out of touch with the ordinary person. Now, let me put to the reader what an architect could say in his or her defence, and then I'll compare poverty excuses. An architect could say, "Well, since I know that there are plenty of people who are happy living in my flats for years, and there is no crime, conflict, or psychological problems at all, the only difference is that these people are well-off; therefore, the problems seem to arise because of poverty. I mean, if the politicians got rid of poverty, then my flats would be perfect for people to live in. I blame all the trouble on politicians and poverty, but certainly not myself or my perfectly designed flats. Therefore, anyone criticising my perfect flats and wanting me to stop building them is nothing more than a bigot using my creations as a scapegoat, to excuse the politicians' failure to eliminate poverty, thereby showing no wish to face reality."

Poverty will always be with us no matter how hard we try to eradicate it, since poverty is relative. What is needed in flats are a lot of twenty-four-hour security guards, which would mean either that the security guards would be underpaid or the poor people living there could not afford them. When I refer to poverty, I mean relative poverty. If you want someone to do labour-intensive work for you, then you have to earn many times the wage that you are paying out to make it worthwhile. A person at the lower end of the economic spectrum could not afford to pay someone else's wages. The fact is that when you have conventional housing and an interracial and multicultural continuity, you can still have a poor but stable and cohesive society without blaming everything on poverty. My own father was brought up in a poor area in the East End of London. He was lucky if there was enough to eat. But that area did not have the crime or the social disorder that we see today. In fact, my grandmother was able to leave her home and go shopping without locking the door, with full confidence that she would not be robbed. I myself live in a country town that is considered one of the poorest in England, but the crime rate is still low compared with the inner-cities of England.

It is impossible to get rid of relative poverty, but it is not impossible to have good housing and racial and cultural continuity. The powers that be should really put their minds to it.

Racism by Misinterpretation, and Understanding How People Express Themselves

Nice People

What is it that makes people nice? Is it education, lack of ignorance, genetics, upbringing? Well, when analysing myself on this matter, I found that some circumstances lead me to behave, as a nice person would not. When I think about crime such as mugging, burglary, rape, and murder, I think up some very nasty things that I would like to do with the "vermin" who commit these crimes.

Some of my thoughts would perhaps shock many an acquaintance of mine. My anger comes from the insecurity and vulnerability I feel, never knowing when I might end up the victim of a crime which could affect my life for many years to come. I have often noticed that an army will inflict punishment on an enemy in proportion to how vulnerable the soldiers think they will be to the enemy in future. If the soldiers believed that the defeated nation or group was going to raise another army or cause more trouble in future, then they might become virtually genocidal (especially if the enemy was of a different racial group). If, however, the victorious army believed that theirs was a decisive victory for all time and that the defeated army or group would accept the situation of defeat and cause no more problems, then the soldiers would show more mercy in aftermath, I believe. People who live in an area of high crime and do not know whom they can trust, and are therefore distrustful of everyone, tend to withdraw into themselves, unwilling to put themselves out or to help anyone else. People who live in a community where there is no crime or hostility but, instead, total security are more likely to behave in a nice manner. Those people who live in a land that is free from internal dissent and hostility are more likely to be at peace with themselves and their neighbours. Unfortunately, if there is internal dissent, then a nasty attitude can arise, leading people to want to "get rid of" such dissent, which means that an otherwise nice people may become rather nasty in conceiving of their means to a solution.

Explaining Racial Differences

On a TV episode of *The Bill,* an ethnic person pointed out to a supposed racist that there are more genetic variations between one white person and another white person (and between one black person and another black person) than there are between black people and white people as a whole. For the reader who is ignorant of genetics, I will elaborate. Just because an organism has different genes does not mean that the organism's structure is different from others of the same type. There is the case of two frogs in the Amazon jungle that have been separated by a greater period of time than the bat and the whale have been separated. The two frogs have genes that are greatly different from one another, − greater than the genetic differences between the bat and the whale. The difference in the genes is not a measure of how different the organism is, or appears. Let's say that a human female egg were split into two, leaving two identical eggs, both female. Add to one of the eggs a single gene that produces the male hormone and changes the otherwise female egg into a developing male egg. The brother and sister that would be the result of such an experiment would be genetically identical, except for one gene. It could then be said that the male was far closer, genetically speaking, to his sister than was any other man on earth. But the physical structure that enabled him to be a man was more like any other man's on the planet than his sister's structure. This proves that the measure of genetic variation is hardly relevant to how individuals differ anatomically or in physical appearance. Most genes between a person of the same race are nuance, genes that create subtle differences between individuals, these genes are very numerous, however the genes between two races are structural genes which are relatively small in number.

When reading about international affairs, I have noticed that when white people are starving, the media have an emotional outburst. However, when it's a coloured race that's in trouble, there is a response, but one that appears to have arisen more from conscience rather than from deep emotions. At one time in Britain, if children were facing extreme poverty, the public would erupt in an emotional outcry. However, nowadays, I sense an air of indifference to child poverty. Is this because we Brits have become dehumanised? Well, let me put my foot in my mouth and say something shocking. Other groups which have immigrated en masse to this country are poor. I would suggest that in the past, if the issue was one of white poverty only, then people's emotions would be aroused and something would be done. But since people of this nation became aware of their own racist feelings, their conscience was stirred and they made up their mind not to discriminate. However, what happened is that if people did not treat coloured people as though they were white. Instead, they simply "dehumanised" white people as though they were coloured. In other words, white idealism is more concerned with the abolition of racism than it is with the abolition of suffering as a whole. This only makes things worse for everyone.

If you hear from news around the world that white people are killing black people, the response is one of outrage. If white people are killing white people, then the response is also outrage. If black people kill white people, then you'll hear the same outage. But if black people kill black people, then the news will hardly mention it. It would appear that people are more concerned about racism than they are about black people's being killed or hurt.

There has been many an accusation of people for revealing racial differences, as though to say that one race is different in a particular aspect is somehow "racist" or "politically incorrect." It does remind me of the Victorian attitude towards sex: if you don't think about it and instead ignore it, then it will go away. There has evolved within the race issue what is referred to as middle-class guilt about race. I must say that I sometimes feel that white people who have a sense of guilt are more often than not of the intelligent middle class. It does appear that people wear guilt about race on their sleeves as though it is proof of their middle-class intelligence. This is a kind of middle-class snobbery, where the middle class can look self-righteously down on the working classes who are expected to do the integrating since it is the working classes that tend to live in the coloured areas.

In Britain, it has come to my notice if anyone wants to express a criticism of ethnic minorities, they say, "I am not a racist, *but* ..." And then a politically correct person says, "Oh yes." Why is it when anyone says anything about race, they often feel the need to say, "I am not a racist, *but* ..."? Perhaps I can answer that question by posing a hypothetical point.

Let's suppose that you are living in the USA during the McCarthy era, when politicians were witch-hunting for communists. Suppose you want to express a belief in having a national health service. You think to yourself how to express your belief. I would suggest that you would start off by saying, "I am not a communist, *but* ..." Or suppose you were in a communist state and believed in free enterprise. Would you not start out by saying, "I am not a fascist, *but* ..."? Or suppose you were living in South Africa and wished to criticise apartheid. Would you not say, perhaps, "I am not a nigger lover, *but* ..."? When a person makes a statement like "I am not a [whatever], *but*...," it tells you more about that individual's society than about the person him- or herself. If you lived in a genuinely broad-minded society, one which does not jump to conclusions about what a person expresses but is willing to hear a person out and take what he or she says as it stands pragmatically, then no one would feel it necessary to say, "I am not a [whatever], *but* ..."

I explain racial differences as follows. First, let's understand a genetic principle: If I were to say that men are taller than women, a feminist might say that this is a myth, given that some women are taller than some men. Well, what I am saying is this: If any man had a cell taken from him and the Y chromosome (which contains the male gene) were removed and his X chromosome duplicated, then that cloned cell would be born a female. That female would be shorter in height as well as physically weaker (and every other "male chauvinist" suggestion about male or female superiority, etc.) than the male from whom she was cloned. The reason why some women are taller than some men is because of non-sexual variables. Therefore, males are taller and physically stronger than females. OK? Get the drift? I hope that this isn't too difficult for the politically correct person or the feminist to understand.

Some PC geneticists have said that when looking at the human genes as a whole, they find no real difference in the genetic structure of people of different races. One could also say that there is not much difference in the gene structure of men and women, either, as geneticists are well aware. It has been found that it is not the XY chromosomes that are responsible for sexuality but a gene which exists mostly on the Y chromosome (but sometimes on the X chromosome). Now, of course, we all know what one gene can do to an individual, not only physically but also intellectually and personality-wise.

Now, here comes the crunch. Let's suppose that biochemists of the future were able to identify all the genes in the human gene pool. They would then be able to identify the genes that are more or less exclusively South Saharan African, those that are exclusively Far Eastern race, and those that are exclusively Caucasoid, in addition to those genes that are common to all humans, of course.

My suggestion is this: Say that there are no racial differences in intellect or personality. If you were to replace, in a South Saharan African's cell, all the genes known to be exclusively South Saharan African with genes known to be exclusively Caucasoid and then you cloned the cell, the resulting organism would be a European Caucasian. I am saying that it is possible that the Caucasian would have a different IQ or personality than the original South Saharan African if the environment were similar. When I say a different IQ, I mean that the person would show a propensity to use inspirational, inventive innovation on the physical universe. Therefore, I am saying that **IF** there is any difference in intelligence or personality, then Caucasians are higher in this type of intelligence than South Saharan Africans or any other racial type. (The difference may be due to a genetic difference in personality rather than intelligence.) If you were to compare two individuals from different races, there are always non-racial genetic variables involved, that could easily contradict any stereotype. Remember, the European lives in a civilisation where the intellectually weak, as well as the intellectually strong, individual can survive and breed. There are quite a few intellectually challenged individuals who's non racial genes who would again, contradict any stereotype

Marketing Factor

In recent years, marketing has been treated as some kind of science, thanks to the input of statistical information, especially with the help of computers. Now, if marketing people are able to prove that 5% of people living in a certain area use a particular product, but only 1 per cent of people in another area use that same product, then it is natural for them to concentrate their advertising in that area of the 5 per cent. They would also choose a shape and colour of product to suit that 5%, as well. This is not bigotry; it is only good business. If advertising is successful because blond-haired children appear in TV adverts, this is only what has been found to sell in the marketplace. (I, as a bald-headed man, do not need to see bald-headed people on TV so I can identify with them; neither do marketers need to sell specifically to bald-headed men. I am not going to stand in the corner and sulk or put my thumb in my mouth, about the fact that bald-headed men generally aren't used as spokespeople. Many people seem to believe that they should be at the centre of the universe and, therefore, that they should be represented in TV adverts.)

Minor Differences in Society

I will now further explain how minor differences, genetic or otherwise, can, in an insidious manner, affect society, especially those differences that have an overall effect on the economy. Let's suppose that the population of Britain were randomly split into two and the government decided to reduce the tax rate for one half of the population by ten pence on the pound. Because of variation in people's wages and also any extra sum of money each person who was part of the chosen half would have, you would not be able to spot which people had the tax reduction by studying any one person, such as finding the total of his weekly pay packet or learning how he or she behaves when shopping. However, due to the laws of economics, the half of the population that received the tax deduction would affect and increase the spending power of the economy. This would be beneficial even to the other half of the population which did not receive the tax deduction, since there would be an upturn in the economy and more jobs available. Therefore, I conclude that a given population with different characteristics, no matter how subtle, can affect a society as a whole. I would suggest that genetics also have a similar effect, especially where intelligence, personality, and mental stability are concerned.

Now for another example. Suppose that there were two races of people who were totally equal in intelligence, that is to say, they both had the same average and variation in intelligence. One race had blue hair and the other race green hair. Each race lived on a separate island. Now, let's suppose that God, just for the fun of it, gave every blue-haired person an extra ten points of IQ, so that the blue-haired race was the most intelligent by an average of ten points overall compared to the green-haired race. Now, of course, the question is, Would anyone be able to tell how intelligent a person is by the colour of his or her hair? Well, of course not, because the normal variation in a race would tend to mask the effect of the individual increase. In just the same way, one is not be able to determine the colour of a person's hair by the results of his or her IQ test. But, of course, the increase in IQ of the blue-haired race would certainly have an effect upon the blue-haired society in that its people would pass more examinations and see their intellectual abilities increase overall. Therefore, the blue-haired people would invent more things and contribute to the pool of intellectual resources at society's disposal.

One could also say that an overall reduction in intelligence would have a similarly dramatic effect, only in the negative. A similar effect on personality, let alone intelligence, could also have a similar effect on a society as a whole. Now, the blue-haired race may object to green-haired immigration into their country, naturally believing that this would lower the national average IQ. When one understands the above principle, one could ask, "Why not discriminate individually, taking into account everyone in the country, so that race does not have to come into play?" Well, the authorities are not going to give IQ or personality tests at the airport. I am not suggesting they should, either. But if a people or society sees a reason for discrimination and cannot officially codify it against certain individuals, then they will discriminate unofficially by using generalisations.

I'll now provide another hypothetical situation in an attempt to explain another principle: Let's suppose that there are two nations. In one of them, all the people have a natural genetic IQ higher than 150. In the other, all the people have a natural genetic IQ below 110. Given how societies are set up, people are needed to sweep the streets and carry away the garbage, just as doctors, lawyers, scientists, and politicians are essential. Therefore, the low-IQ society would have its doctors, lawyers, et al., just as the high-IQ society would have its road sweepers et al. The professional people in the low-IQ nation would be well educated and see an increase in their IQ because of the extra stimulus. Compare this to the non professional people of the High IQ society, who would be the least educated and have their intelligence dulled by the monotony of their work. Now, if a professional from the low-IQ society were to visit the high-IQ society, he or she could well look down her nose at a road sweeper who lives in the high-IQ society, expressing an attitude of,

"I am a doctor; therefore, I am intellectually superior to you." I believe the reader can see that one's profession does not proof one's intellectual superiority than being of a particular race.

Two Worlds

The strange thing about the human population throughout the world is that it seems to be split into two worlds (barring the ex-communist states): the Western world and the Third World. Why is this? Well, let's investigate. It could be said that geography is the cause. It is true that Third World counties are in the south, while Western countries are in the north. What, then, about Australia and New Zealand? Aren't Australia and New Zealand in the southern hemisphere and the landscape barren to begin with?

One could say that Australia was more productive after the arrival of Western people. There was a surprising rise in population of kangaroos and other adaptable wildlife ever since white people arrived. It would be no exaggeration to say that white people turn a barren wilderness into paradise and that Third World people turn paradise into a wilderness. Brazil has far more natural resources per head of population than Australia, and also a higher population as a pool of human resources. Many would say that it is depends upon education. Yes, fair point, but the British and Irish prisoners who first settled in Australia were hardly literate middle-class intellectuals.

They had to have the willingness to learn and the willingness to act upon their new found knowledge. They also needed opportunities. If a cool climate is what turns a people "culturally" into Western people, then why are not the southern Argentinians or southern Chileans a Western society? (Argentinians might consider themselves to be a Western society but then they are racially close to Southern Europeans). It would appear that the only real difference between a Western society and a Third World one is race and only race. Of course, it is true that dictatorships encourage poverty, but the Soviet Union was the greatest oppressive dictatorship on this earth but still managed to maintain certain standards. Even with the turmoil that ensued after the fall of communism, Russia still manages to get by – well OK just. Also, it is true that there are very successful coloured people in the world. This is possible because of two main reasons.

The first reason is this: There are many ways a person can be successful. One may be a successful musician, actor, buyer and seller of goods, doctor, solicitor, pop singer, or artist. None of these things in itself goes to create a Western society. These talents abound in the Third World, anyway, so what's the big deal? There are many people in the Third World who succeed at these endeavours, but they only make themselves, not the society as a whole, successful. It is no surprise that coloured people with these talents are even more successful in the Western world since there is more wealth around to call upon. To give an example: When Asian immigrants came to Britain, they found a shortage of shops selling Indian food (surprise, surprise). With plenty of other immigrants from the same culture, people figured that there would be a demand for said shops. As many Indians are used to being shopkeepers, it does not stretch the imagination to determine that they had a ready-made captive market which the "home team" knew nothing about. Therefore, they only have to compete with people from their own culture for customers. The number of shops owned by the native population in Great Britain is decreasing because of the revolution in out-of-town shopping centres, or hypermarkets. We shall see what happens when the hypermarkets start to sell a wide selection of Indian food! Also, when it comes to manufacturing, immigrants tend to exploit their own people by employing them in "sweatshops," which most white people in Great Britain would not even attempt because of the many conditional legalities (which many immigrants have been known to ignore). Any odd exceptions to the point I made in the last paragraph do not devalue the point as a whole. Also, some ethnic pop groups are very successful, for instance the Jackson 5 in the United States. Michael Jackson was a multimillionaire because he was talented in singing, dancing, music writing. (Although a good business man I have been told, there are plenty of excellent business man in the third world but this talent does not turn a Third World country into a western world). These talents, although they bring a great deal of pleasure to many people, me included, also exist in the Third World and would not in themselves contribute to making a Third World country as financially successful as a Western one.

The second reason is as follows: If a man in the Third World only competes with people of his own racial type and if people of his race have low abilities in a particular field, then the man who performs best will be successful. He will also gain knowledge, wealth, and influence in his own country, which can then give him an edge when he competes in a Western society, where he has more wealth than the average Westerner. He will then project himself as more successful. I am, of course, generalising. I would point out that it is an accepted fact that the doctors, nurses, and scientists who come to Great Britain from Third World countries are not necessarily equal in standard to their British-trained counterparts. (However from personal experience this is not always the case).

One reason given for the poverty of the Third World is the Third World Debt. It so happens that after the Second World War, Britain was up to its neck in debt from protecting the world from Nazism. It so happens that Britain has paid many times more in interest than the original amount borrowed, but you don't hear British people bellyaching about it to the world, as though it were some poor victim of the United States.

While we are on this subject, I have heard that some Americans boast that were it not for the United States, Britain would be a Nazi state. Well, no doubt. But if it were not for Britain keeping its *giant aircraft carrier* afloat, the Americans would not have been able to conquer Germany, either. This would mean that the Germans would have had plenty of time to eventually develop the atom bomb and the V-3 rocket.

If those had reached the United States, then America might have ended up a Nazi state, as well.

The Asian Myth

I remember hearing on the radio many years ago that Asian children did better in school than their white classmates. I must admit that I was surprised to hear this, retaining a certain disbelief. I began to think it over and remembered that it did not say, "Asians did better than whites, full stop, but that they were better than their *classmates*. It then struck me, after a couple of minutes, that the findings came from an ethnic area. Most Asians in that area were of middle-class origin and had high aspirations, but the whites of that area were of working-class origin and had low aspirations. The fact is that Asian parents are more likely to be aware of their ethnic status and know that their children may be discriminated against in the labour market. Asians are prideful in not wishing to appear less than white people. The highest-performing Asian children were being compared with the lowest-performing white children. I reached this conclusion long before the whole truth came out in the media. The people doing the research had reluctantly given out the data before proper analysis was done.

I have also noticed that many mathematics examinations ask questions about today's economy rather than about straight arithmetic. Therefore, the questions measure memory rather than mathematical intelligence. If the whole of the present-day curriculum were made to suit a student's ability to remember text, then the white girls (and many ethnic children) would certainly, on average, outperform the white boys on examinations. Is it so surprising that ethnics and females are coming out on top in these subjects? After full analysis of the above-mentioned data was achieved, which I assumed was more or less correct, it pointed out that most Asians, on average, stayed on in school an extra year than whites so as to achieve this result. To be fair, I would point out that some remarkable mathematicians are of Asian descent. I believe that this is because Arabs and Asians, engaging in trade at the crossroads of the world, developed the ability to calculate in the head the value of goods. Also, white lads tend to be late developers. Their brains do not fully develop until they are in their early twenties.

Regardless of my *apparent* racist views, (Physical Anthropology is about the physical attributes of humankind, I do not see how one can ignore race as an issue of this science.) I am not ready to chuck anybody into the gas chamber, at least not for racial reasons. The decision to exterminate human beings is a point of humanity, i.e. love and hate, not a point of who is capable of this and that – or not. If one has enough power and wants to exterminate anyone, one hardly needs an excuse, although most people who murder others on account of race try to find one. I think it's about time that we sorted out in our own minds whether we are talking hatred of a different group's personality, hatred of their physical appearance, or an awareness of difference in ability in which we honestly believe, given the circumstantial evidence available.

Race and Culture

In the first place, it could be said that the environment influences behaviour. People behave in a certain way in order to survive in a particular environment. But it is obvious that if people's inner characteristics naturally complemented the environment, then they would better survive. An example is seen in the tendency of people who live in northern climates to store away food and material for the winter. We all know of people who build up a stock of food for the winter, preserving jam or pickle chutney, even when it is unnecessary to do so in civilised times. But in times past, behaving this way would have meant the difference between life and death. Many of us are procrastinators, not wanting to do something which we know needs to be done because we do not like doing it. Some tend to put it off until tomorrow, but, of course, tomorrow never comes. In some cases, people put something off until the last minute and then panic because it may be too late to do things properly. It is those people who store goods away like a squirrel who would have, in times past, survived the best. Although storing goods and "making hay while the sun shines" is part of the northern culture, the human desire to do this has evolved over the years as an inherited instinct does. It is evident in some people more than others.

1. Some people will store away for the joy of it, even though there is no need to.
2. Some people will *happily* store away, but only when they know that the goods will be necessary in the long term.
3. Some people will store away reluctantly, approaching the task as a chore when they feel they have to, but will only store as much as they think they need.
4. Some people know that they should store away for the winter but make every excuse not to do so until the last minute – and then they may find themselves doing things too late.

It is quite obvious that it is not simply culture or necessity that makes people the way they are. Also involved is the nature of a person who is "happy" to conform to cultural necessity. It is quite obvious which people of the four types mentioned above would best survive the worst environment. Also, if a person has a surplus during a bad winter, he or she could use that surplus to barter with those who have no such thing. The extra income would give the person who over stored food greater status as well as a greater chance for his or her family's survival. If a really bad winter occurs, then he or she will last out the winter, whereas another would starve or die from an ailment that affects a weakened immune system.

It is like this: The environment not only shapes a culture in the short term, out of necessity, but also shapes people's genes in the long term. Once the genes have evolved to suit an environment, the people in question, when put in a different environment, would still want to behave in a similar manner. However, if the above-mentioned people were adopted at birth and raised in a different culture, they may initially conform to the behaviour of the adopted culture but still show signs of wanting to store away. Their parents may positively deter them from doing so. When the first adopted generation reached maturity, they might, out of cultural habit, not store away, but their children may retain an inclination to do so. They may be undeterred by their true parents, who may appreciate the naturalness of their behaviour. When it comes to culture, I am of the opinion that it is the genetics of a people that creates and shapes the culture in the long term (which may be many hundreds of years), but it is the culture the shapes the individual in the short term. I am fully aware that ethnic people who were brought up in Britain and identify themselves as British are indistinguishable, for all intents and purposes, from native British people in their behaviour. However, when ethnic English people get together, they tend to develop their own culture with their own distinct mannerisms. In some cases, the only cultural similarity between native English children and the children of immigrants is the accent.

Here, I provide a hypothetical situation as an example. Let's suppose a great number of white Western children were taken from their parents at birth and were then brought up in the Third World. At maturity, they were all put on an island similar in climate to their land of birth, away from their adopted parents and culture. I would suggest that, at the outset, the first generation would be more like their adopted parents in behaviour. After many generations, especially once the young people of each generation attempted to rebel against the older generation, the culture would lean gradually, generation by generation, towards Western progressiveness, even if the language and mannerisms were those of the Third World country.

To give an example: There are jokes about the Welsh. "The Welsh are OK, so long as they are led by a white man." Stan Gooch has said that the Welsh are Neanderthals, relatively speaking. I would suggest that the Welsh are a Western people who are influenced by Indo-Arabic culture. This is probably because, originally, a tribe of seamen from North Africa settled down somewhere in Wales. (It has been suggested that the Welsh are culturally North African in origin.) As more Western people joined them, they took on their language and culture until, racially speaking, they ended up as mixed Northern Europeans just like people of any other European nation, but they showed some of the cultural characteristics of a North African culture. If you listen to the Welsh accent, you will notice that it is a Peter Sellers–type Indo-Arabic accent, except that with the Indo-Arab accent, the words and sentences finish with a downwards-descending tone, whereas the Welsh accent is upwards-ascending. I suggest that it is the genetic nature of the white western Welsh that has influenced their Indo-Arabic speech. Recently, I heard on the news that white Mohammedans (who have been influenced by Arabian culture) of Bosnia around the capital city refused the available food because their own people were starving elsewhere. I have never known the Arabs to be quite so self-sacrificing in this way. Also, white Mohammedans have been heard to say, "We are not all that religious, but we follow the Muslim culture." This sounds very European to me. I will give a more detailed summary of race later.

More Explanations of Racism, with a Summary

There is another form of racism that I haven't yet mentioned: slavery-supporting racism. Some white people feel a need for blacks to exist so that they can dominate them, enjoying having someone in an inferior position (regardless of whether one believes them to be inferior or not) so that one can feel superior. (I will delve into this in detail further on.) This is different from the type of racism I already mentioned, where one simply wishes to be away from black (or coloured) people but has no wish to hurt them physically or mentally. Racists of this type have only an intolerance for being close to people of other races (racism by avoidance). The point here is that the white person has no evil intent or malice towards the black or coloured person, only a desire to live with white people only (again, regardless of whether the white person believes black people to be inferior or not).

Of course, many white people may be guilty of both types of racism to a greater or lesser degree. Whether it's to do with race or not, I would say that we should think carefully on how we treat other people. One should practise having good manners, as it is so easy to confuse the difference between racial feelings and bad manners. A white person should not treat minorities with bad manners, but white people should not be made to carry a sense of guilt when they exhibit a natural desire to live within a predominantly white community.

I remember a instance of an actual experiment in the United States which was portrayed in a film of name I cannot remember. Pupils were given armbands which classified them according to the circumstances of minority groups in the area. The restrictions and the degradation became intolerable for the white students, which is of no surprise. One white lad with an armband representative of a low class was at the end of his tether. He stood up and said, "It's not my fault that I was given this low-status armband."

A person of a minority group stood up and said angrily, "It's not my fault that I was born in a ghetto. It's not my fault that my mother has to go out to work as a sweated labourer to keep me and my brothers and sisters." Well, good point, *but* why did his parents have so many children? Wouldn't this have been a contributing point? (In this case, the people involved immigrants – and probably illegal immigrants from a Third World society which they left on their own accord. They can hardly blame white people for their poverty, since they were poor to begin with, with or without white people's help.) No one asked that boy's parents to have so many children – or even to marry in the first place. Many people in the Western world refuse to marry or have children because they fear poverty. (Of course, I do not recommend exploiting minority groups. Still, many white people are so frustrated by minority overpopulation that they will pay minorities lower wages. The reason or excuse for this is that no matter how much you pay minorities, they will always be poor, because having big families is simply their way.) I would now suggest a follow-up argument. (My creation only)

Ethnic man: "Oh, but we are Catholic. It is not our fault that we have large families." [This person uses religion to evade responsibility for his or her actions.]

White man: "Oh well. Maybe a fascist was brought up to be that way, but that does that excuse him of all social responsibility for his actions. Remember, religion is only an idea."

Ethnic man: "Oh, but that is a bad idea."

White man: "Well, who says that having a large family that society has to help support is a good idea?"

Returning to the topic of the Ethnic student quoted above, I know it can be said that it is not his fault that his parents had a big family. But the odds are that he, being a "good Catholic", will marry and have a large family of his own and therefore perpetuate the problem. He cannot simply throw his hands in the air and avoid all responsibility when his own children will end up in the same situation. If the son commits the same sins as the father, then he must take full responsibility for his own predicament and not blame society for all his troubles.

The strange thing is that although left-wing people often talk about overpopulation and the sanctity of planet Earth, they seem ready to protect people's "natural" tendency to breed out of control. I will make my point about naturalness a little further on. It is said that they who live by the sword die by the sword. It could equally be said that they who live like the rabbit will die like the rabbit: by way of deprivation, predation, (disease), or starvation. Yes, I know I sound hard, but I think that white people should treat people of the Third World with a bit of tough love – at least those who do not take responsibility for that over which they have control. I believe that many an individual who was born in the Third World or is part of a minority group and who has no control over his or her environment would subscribe to the Western way of life. Doing so may not stop him or her from living a life of misery since he or she is still part of a society, over which an individual has no control. This type of person does have my sincere and total sympathy, as his or her circumstances are the greatest injustice. I wish to repeat that *I would like people to take responsibility for that over which they have control.* I do not believe that people should be held responsible for anything that they do not control. If you think that I am a fascist, know that I am only too aware of "There but for the grace of God [or different circumstances] go I.

Here, I would like to mention a passing thought I had about reincarnation, in which I do not necessarily believe. I have wondered about near death experiences, which point to the existence of an independent soul. If we all have independent souls, then all we create on this earth is the carcass in which the soul resides. If so, do we not have a responsibility to produce the best possible Human bodies for future generations to exist in, so that they can live full, productive, and happy lives? I mean, just as we have the responsibility to produce safe, reliable cars for people to travel in, we have a responsibility to produce reliable Human bodies for future souls to inhabit and travel through life. It is only humankind's conceit about producing its own kind that has allowed the pathetic variety of Human bodies that we see about us – (and in the mirror) – to develop. I would suggest that if we were to practise genetic selection, then we could produce better Human bodies for future souls to travel in. Who knows, it might be you or me in a future life. I would hope that good physical and mental health would be the criterion for selection. More on this later.

I have often wondered about the principle of putting oneself in someone else's shoes. Is this a principle that only white people should be aware of? Let's suppose that someone like Martin Luther King had had a body transplant as a baby, that is to say, his brain was put into a white baby's body, and then that baby was adopted by a white couple. So, he grew up with the belief that he was all white. Would he then be a great campaigner for equal rights in the name of the brotherhood of humankind? Would he make a speech like "I Have a Dream"? I am not saying that such a thing is impossible, since white idealists do exist, but somehow I doubt that he would. It is just as possible that he would grow up with all the prejudices of a white honky.

An idealist should hold views for purely idealistic reasons, not for reasons of self-interest. Let's suppose that a wealthy man claimed to believe in equal taxation by percentage and, therefore, opposed an upper rate of tax. Would you consider his belief that of a noble and idealistic man? I would hardly think so, since it is only in his self-interest to hold this belief. Should we then think of a black man who believes in racial equality as being self-sacrificing and idealistic? Again, I hardly think so. The point is that white people are often persuaded to consider coloured people's points of view, but a coloured person is rarely asked to put him- or herself in a white person's shoes.

The reader could ask me how I would feel if I had a black body transplant as a baby and was brought up as a black child in a white society. Well, I must say that I would be pretty pissed off when I saw the prejudice around me. Perhaps I would resent of my lot in life. But what right have I to insist that people should love or even like me? No one can alter what oneself or anyone else feels. Even though I am a white man, it is possible that many people may not like me. So, what? The world does not owe me or anyone else eternal happiness. All we have, is the right to be left alone and not to be victimised. The right to be left alone is, I think, the first right of all humankind, not the right to be liked, loved, or anything else. If the politically correct continued to concentrate on this and less on the imposition on the white person's "personal space," then they might get greater support from the white community. The fact is that we are all discriminated against, especially those who are relatively hard up and haven't the power over their own lives that money brings. As a white man, I do, at least, try to put myself in a black man's shoes when I happen to meet one. I must remember not to patronise or show a disdainful indifference, as this is evidence of a bad attitude.

I remember many years ago seeing a documentary (I cannot remember the name of the person or the name of the documentary) on Third World poverty. One commentator said, "Just because they have lived in poverty for most of their lives, it does not mean that they are immune to human feelings. The death of a child is just as bad to them as to any Western person. It certainly is not for us to judge their feelings – or anybody else's, for that matter." I certainly would not argue with that. However, suppose you were to say to people of the Third World that many Western people do not marry or have children, or else they limit the number of children because they would be poorer if they had more. What would we say if members of the Third World were to say (and I believe it has been said), "With you Western people, it's all right for you to remain single and/or childless because you do not feel the way we do about family and children. You have no feelings like we have. I mean, if you had the desire for marriage and children, then you would have married and had children, wouldn't you have?" One could say that if you do not like to be poor, then you would not make things worse by having more children. Perhaps you don't mind being poor?

I know that in many circumstances, being single and childless does not guarantee immunity from poverty, no matter where you live. Also, if a person of the Third World did everything required by a Western person such as remain single and childless, there is again no guarantee of freedom from poverty. We can all be the victims of the society to which we belong. One cannot always choose one's nationality, and one can never choose one's race. The main point I am making is about society. There will always be people locked into a society not of their choosing. Some people are willing to live in another society, to which they would be quite willing to adapt. But it is very easy for a Third World person to claim that he or she is "white" (Western-minded) underneath the skin in order to gain entry to the West. Then, after becoming a citizen of a Western country, the person may cry, "Human rights! We have a right to breed." How do we Westerners discern when awarding citizenship to people from the Third World? Do we have the right to discriminate?

Character

Allow me to indulge in a bit of analysis. (Please note: this is my own personal analysis.) I would say that character is evidenced of the higher brain function's ability to control or override the lower brain function. I shall give an example. Say that an individual drinks alcohol, uses drugs, and gambles, which he consciously knows is not good for him. He says that he cannot help himself. It is not that he needs to ask someone's permission to stop or that someone is forcing him to indulge. The problem is literally in the person's own hands. It is the strength of his upper brain over his lower brain that regulates a person's behaviour. This, I would say, is strength of character.

While we are on the subject of analysing words that people use to "blow their own trumpets," let's analyse *personality.* I have used this word in the context of an instinct or a desire to do or to be. But now I wish to define the term to reflect the way we judge people by their personalities – when we say in the social sense that someone has a strong personality.

Personality

Personality is not so much how people control themselves, but how they are able to "control," or seduce and persuade, other people who are in awe of them.

Boxes

I have known many people who have said, "One should never put people in boxes, as people are all individuals." On the whole, I would agree with this, but is it not true that people tend to put themselves in boxes? Surely, when one sells one's soul to a religion, one puts oneself into a box, believing what everybody else who subscribes to that religion believes. Although there are many offshoots of religion, very few individuals will go it alone. Where is the individuality? Also, human characteristics tend to be put into boxes (but this only refers to the characteristic, not the whole person). In other words, when a person refers to a particular human characteristic, it does not mean that people are either this way or that, only that a person in whom that characteristic is predominant may, on average, tend towards this way or that way of thinking or doing. Whenever those people who hate putting people in boxes become aware of racism, they almost inevitably talk about the racist as though he or she were a member of a fundamentalist religion and had a singular point of view on all things. I believe that those people who voted for a racist councillor in the East End of London probably held different political views, far more varied than those of the people who would condemn them.

Racial Summary

The first thing about race or any group of people is the behaviour as a group, in contrast to individual behaviour. The greatest delusion of civilised people is to believe that societies are equal to the sum of their individuals. It is a known fact that two men can carry one piano much more easily than two men can carry half a piano each. There is always the contradiction that many hands make light work and too many cooks spoil the broth. I believe that a people can be greater or lesser than the sum of the individuals. I remember reading that a Greek philosopher losing his temper with a crowd of people, asking how they could be so intelligent as individuals but so stupid when in a group.

In order to make a point about race, I will separate the world into several cultures. Readers can make up their own minds about whether a pattern of events is involved. Knowing the racial content of a nation means that you can predict in the long term approximately what the state of that nation will be.

First: The Rich World

This world is filled mostly with rich people who enjoy a good standard of living. The people can be hard-working when good incentives are involved but lazy when things get too easy. Still, they can be very hard-working and at their best in wartime. They are very inventive and innovative and have built up a civilisation on hard work, with an eye towards automation in order to save having to work hard in future, rather than working hard for its own sake.

Second: The Hard-Working World

People of this world may be rich or poor, but they are very industrious and hard-working. They are known to live to work rather than work to live. They are not very "inspirationally" inventive but can make the best of other people's ideas, perfecting every introduced idea to its optimum. These people may live in a highly automated or very simple society, but, either way, they are dedicated to their society. They are willing to sacrifice their standard of living and their own individuality to engage in social cooperation for the common good of their nation.

Third: The Overpopulated Poor World with Very Few but Very Rich People

This world contains a great deal of human misery amongst great wealth. It produces a few inventors, but very few of their ideas are realised because the society has an unstable structure and excitable citizens. The poor majority are "thought" to be rather lazy and apathetic in spite of their excitability. Some (shopkeepers, et al.) do well at trading but not at producing. Production is normally realised by cheap, sweated, "forced" labour rather than automation. The people's religious devotion seems to keep them overpopulated and their minds inhibited. Therefore, they tend to wallow in poverty and love of projecting themselves as being victims.

Fourth: The Very Poor World

This world always seems to be poor. Its people are uneducated and its governments corrupt. While there may be a few wealthy people, these are mostly in government. The people are generally "thought" to be rather lazy but will work hard when they feel they have to. They are not thought of as particularly intellectual in nature, even though some may be well educated. But even those who are well educated and have acquired an intellectual air about them are not known in general to come up with any breakthroughs in technological development. They are known to be courageous in battle but rather ill-disciplined. They make good solders with the right leadership. The very poor live mainly on the edge of starvation. Nations of this Fourth World need help in some form or another, although they can get by OK if they live within their means and are able to understand their own strengths and weaknesses.

Analysis

If you haven't already guessed, the First World consists of the white race of people, predominantly made up of Teutonics with and an admixture of the Mediterranean race. The Second World is the world of the Far East Race. The Third World consists of the dark Caucasian or Southern Asian, Arabic, and Latin American gene pool. (Southern Europeans could be included in this group, but to a far lesser degree, since there are more Northern European elements in that gene pool.) The Fourth World is the world of the South Saharan African race.

I would point out, however, that in the First World, society of pure Nordic peoples tend to have a higher standard of living than the more Southern Europeans, they also tend to lack the "wish fulfilment" factor, as I mentioned earlier in this book. It is because of a small proportion of Third World elements (genes that exist throughout the Third World and which flow through the Western world's gene pool), which have existed in Northern Europe for hundreds of years, that the "wish fulfilment" factor drives the First World element to invent a fulfilment for the Third World element. I will say again that the Nordic race never became civilised in and of itself. It only did so after mixing with or in contact with, Third World elements.

I would point out here that nations that are mostly Nordic in type which would include Iceland, a baron island in the North Atlantic, tend to have a higher standard of living than countries like England and France which are far from pure Nordic.

Every nation's prosperity and success tends to ebb and flow along a margin relative to the rest of the world, so it is not surprising when a First World country sinks to a level that a Third World country would be lucky to rise to. This happens from time to time, but only when the Third World country's population is not too different in its racial elements from that of the Western country's population.

Unless there are some exceptionally bizarre circumstances like in Kuwait or the former Soviet Union, there do not seem to be any exceptions, as far as any society as a whole is concerned. Even the Russians under the yoke of communism made some revolutionary inventions and developments. The Kuwaitis do not seem to have any great success in research and development even though they are able to pay for the very best education for their children and their people. However the people they employ from outside, although not Kuwaiti citizens are a part of the economic community. If they did not have imported labour they would have to get their own rich people to do the dirty jobs. Therefore can you imagine rich Kuwaitis sweeping the streets? The question is how many people do they employ per head of population and how much do they pay this imported local labour?

Remember, I am referring to a comparison of societies, not individuals or groups whose race originated in one society and who are now living in a foreign society as a small minority. As I mentioned before, even if everyone in a particular nation ran a corner shop or made clothes in a sweatshop (manufacturing goods by using cheap labour rather than efficient automation), no matter how successfully, this would not make that nation a successful First World society.

When it comes to a progressive culture, I would say that some people's ideas helped create the human culture we enjoy today. Then, there are people who read, absorb, and memorise information but do not necessarily add to the sum of knowledge other than by performing routine laboratory experiments. There are also people who may absorb the accumulated knowledge of the past and then add to that knowledge by engaging in deep thought and meditation. Just because non-Western people can absorb Western knowledge does not prove that they are capable of making continuous Western-type progress by themselves and in isolation from Westerners. Holding an academic degree in one subject or another only proves that one is capable of absorbing knowledge through long, hard study, not that one can create new ideas. Although I am not suggesting this is impossible for a specific third world person.

Aspects of Society

I remember one psychologist's talking about James Bulger's killers. The psychologist pointed out that if the two boys convicted of the murder had been with other, normal children, then they would probably have learnt that one simply doesn't commit murder. But because they were two psychos together, they psyched each other up. It is not that they are the same in all characteristics; it is just that they have a characteristic in common. Of course this is a negative example of society, but there can be very positive examples of the same principle. It is this that makes society the way it is. What makes a nation different is not that the people are all the same, but that the people have a characteristic in common.

Defining the Different Types of Racism: Jealousy, Racial Feelings, and Intolerance

Jealousy

Many people misunderstand the concept of jealousy. *Jealousy* is one of the most misused words in the English language, as it is generally used to excuse any detriment found in or insult directed against oneself. As an example, if a child comes home crying to her mother about some insult she received, the mother will almost inevitably say, "Never mind, dear. They are just jealous." This may make the child feel better temporarily, but when the insults continue and the child is not happy about her own position or characteristics which inspired the insult, her mother's saying that the offender is jealous does not wash very well.

Examples of Jealousy

Let's suppose you are a man attending a party where there is a beautiful woman on whom you've had your eye on. Into the room comes another man who is very good-looking, tall, healthy, decent, and honest – everything you would like to be yourself. At this party is your single, unattached sister, in whom the stranger takes an interest. In such circumstances, you would normally be quite pleased. However, if the stranger showed a great deal of interest in the woman in whom you are interested, then you would be somewhat miffed. This would be a case of pure jealousy. If, however, the stranger remained with your sister, you would later be pleased if she married him and the two had many children (your nieces and nephews).

Suppose, now, that the stranger has the appearance of Quasimodo – a repulsive individual (as a perception of course) – and also has the ability to seduce a woman into thinking that to reject him would be unkind, selfish, and – yes! – bigoted. Perhaps your sister identifies with him as an oppressed individual, seeing herself as being in the same position, albeit for different reasons. You cannot blame yourself for wondering what sort of nephews and nieces the couple would produce, especially if the man's deformities were genetic, in which case your sister would bear semi-Quasimodo-looking children (your nieces and nephews). An Ugly individual in this instance is what is perceived as Ugly, although it can still be perceived instinctively and emotionally as a deformity. However suppose this Quasimodo took an interest in the girl you had your eye on. You might feel what a waste of a beautiful girl. Remember, this last instance is not same sort of jealousy, unlike the instance mentioned in the prior paragraph.

You can see that jealousy is not the simple thing that some people make it out to be. To accuse a person of jealousy, I believe, is a very underhanded way of attacking someone who made a criticism. It puts the critic in the position of having to spend time to defend him- or herself against something that cannot be proved one way or the other, since no one can read another's mind. Therefore, no proof or evidence would or could be put forward to prove that jealousy does not exist. In this way, the criticised person can avoid any forthcoming criticisms.

Another point could be made this way: If I saw a film wherein Tom Cruise's character made love to a young, beautiful woman, I would feel no sense of revulsion. However, if I saw a ninety-year-old man making love to a young woman, I would find this most obnoxious and in bad taste, the same as if I saw Esmerelda being made love to by Quasimodo. According to political correctness, I must therefore believe that I am jealous of the ninety-year-old man and Quasimodo, but not of Tom Cruise.

In summary, I would say that jealousy, as it is generally understood, entails envy along with resentment. I would say that people tend to feel jealous when an individual or group has something desirable but is seen as undeserving of it. Some people feel that the haves acquire things underhandedly, which is perceived as unfair. If a man wins the lottery or the pools, then he has won a game. All who lost the game knew the rules and lost by fair means after willingly paying a stake.

However, there are instances when individuals earn huge amounts of money by speculating on the value of the pound, such as the time when the pound was devalued and a certain individual earned a few billion by speculating, although the devaluation was partly caused by the speculation. This is not a game the British public can join in, and not one they would particularly agree to, although they know that it exists. This instance of speculation lost the country a lot of money. The general public suffered in the long term, although they were not involved in and didn't agree to pay any personal stake. You can see now why there is more resentment against speculators than against lottery winners.

Racial Feelings

Racial feelings have to do with how one may feel about people of a different race – how one may respond to their physical appearance or the mannerisms of their culture. Of course, culture influences behaviour a great deal, but the culture of a people does, in the long term (over many generations), reflect the overall natural tendency of a people. More on that later. It is possible to dislike a race of people without believing them to be inferior.

To reiterate what I already have said.
If a person does not like anything that, that surrounds them "one gets over it" by keeping what one does not like away from oneself. One might believe, rightly or wrongly that one has a right to isolate oneself from unlikeable situations. Just as one would not have the right to expect someone to go into the cold without a coat, it would therefore be wrong to deny the right of a person to wear a coat.

If a man wants to buy a house and has two choices, either living in an all-white estate or a mixed-race estate, and he chooses the white estate for racial reasons, then would the law or any power try to stop him and force him not to discriminate? Of course not. But if too many other people make that sort of decision, then the all-white estates will be more in demand than the mixed-race estates; therefore, the price of the all-white-estate houses will go up and, relatively speaking, the price of the mixed-race-estate houses will go down.

Many things in life can go wrong and be detrimental to a person. One could have an accident, fall ill, divorce, or even go deaf or blind. One can theoretically insure oneself against some of these things in the event that they cause financial loss, but not against the devaluation of one's house if ethnic people move in, whether or not one minds living in an ethnic area. When all is said and done, I am aware that this may seem unfair to black people and also how I would feel in their place. The amazing thing is why would ethnic people want to buy a house in a white area in the first place and live amongst white people, not other ethnic people (his or her "own kind"). I have often wondered whether black people like to live amongst black people. It would be ironic if it turned out that black people were just as racist as some of them accuse white people of being. It always seems so strange to me when a black person says how proud he or she is to be black, that black is beautiful, and that white people are wicked, but then wants to live amongst white people, with all the hassle and hostility that it may bring. I have often wondered whether black people like other black people as much as they appear to on the surface.

Black people do not seem to have the social unity that all their brotherhood and sisterhood jive would otherwise justify. Most violence is black on black, not black on white or white on black. Unless we all face the objective and subjective realities of Black not liking Black, I feel that there will be no answer to the problem. My mind goes round and round, trying to make sense of things and find a true answer. But to suggest that white people are not entitled to have peace of mind is, I think, the greatest sin.

The point is that so long as the media is controlled by the "guilt" of the liberal-minded middle-class white person, who seems to have lost all confidence in his or her sense of right and wrong, and only discusses race when it is to ridicule other points of view, the race issue in England or any other country will never come to any successful conclusion. Anti-Racism should be put in the same category as anti-greed. No one is suggesting that racism is good, only that we should do everything we can to take the sting out of it.

Fear, Ignorance, and Political Correctness

In today's world, we are surrounded by technology that very few of us understand. One might ask how many doctors are able to understand silicon chips, nuclear physics, rocket science, etc., or how many rocket scientists understand silicon chip technology. You can see that everybody is ignorant about much of the world. Even middle-class professionals are aware of their own ignorance.

It could be said, "We are all ignorant, only on different subjects." We could all be self-conscious about our ignorance. Now, *ignorance* is a word of abuse, in the sense that an ignorant person is lowly and unworthy. (Calling a person ignorant is like calling a black person a nigger. Both are abusive terms.) Ignorance should only be denigrated when one wallows in ignorance as though it is a virtue..

I believe political correctness feeds off of fear and ignorance. The PC person harps on about racism being caused by ignorance, saying that all white people who express a dislike for coloured immigration are racist and therefore lowly and unworthy. Politically correct people are able to create a fear that one's own society will look down on one as ignorant, if one does not conform to political correctness. This is like the story "The Emperor's New Clothes," which describes a whole population of people who are duped into believing a falsehood because they fear looking stupid and ignorant. If people had confidence in their own intelligence and knowledge, then this phenomenon would not exist.

The In-Crowd

I believe we all know of the in-crowd. When at school or, perhaps, at work, one would sometimes find a group of people, perhaps good-looking, who were very popular and exciting and often laughing. Most everyone wanted to join them. But they were found to be exclusive, only allowing a few people into their group. It may have come about that those who were rejected by the in-crowd felt somewhat put off and offended, taking the rejection personally. The in-crowd may not have wished to give offence and may have claimed that it was nothing personal. The general reaction of someone who took the rejection personally might be, "Who wants to join your *silly* group, anyway? So, there." The rejected people then ganged together in a cauldron of mutual enmity. The strange thing is that if any person amongst the rejected group had been accepted by the in-crowd, then he or she would probably have been willing to reject others, just like the rest of the in-crowd. This scenario may explain why many minority groups hate white people, because minorities may have experienced discrimination in the workplace and so forth. Hell has no fury like a person scorned. This could include a white person.

Let's analyse: When immigration to Great Britain began, West Indians arrived in a state of innocence, quite happy to integrate and become natives. However, they found that the native population rejected them ("You can't be a part of our gang"). I would say that this was not necessarily unnatural. One cannot blame the native population for wanting to be just themselves, to go on living as they were used to living (in the sociological sense). However, one can't altogether blame the immigrants for feeling resentful. They and their children turned inwards, identifying themselves as not a part of society.

When an ethnic person sees a crime committed against a white person by another ethnic person (or a non-ethnic person, for that matter), he or she might say to him or herself, "Why should I do anything for that white victim, since I am a non-person? That will teach them." Since serving as witnesses and being alert for criminals is on the front line against crime, the native population might believe that ethnic people are not helping to keep law and order. Since law and order does depend on everybody's playing their part, a white person may see ethnic people as helping the criminals. "All ethnics are all criminals" could be the cry. "They are all in league with the criminals, if not criminals themselves." Give a dog a bad name and the dog gets kicked and then bites back. Then, people say, "There you are. I said it was vicious." You can see how a little wrong can become a bigger wrong and then get out of control. I would like to point out that no matter how much one believes that people should integrate, people have a right to be themselves without having to face the threat of crime to their way of life.

Everyone must learn that when they go to another country, they cannot expect a bed of roses any more than the English emigrants in Australia should expect a bed of roses over there either. They should not expect the "home team" to fall all over them with welcoming arms. Of course, not all ethnic people in predominantly white countries are immigrants. Some are the sons and daughters of immigrants, but unfortunately for them, the same principle applies. A free society, which many an immigrant came to Great Britain to enjoy, has its sharp end, at least from the non-white immigrants' point of view. The white population is expected to be tolerant even if they don't like mass immigration. Minority groups should also show a little tolerance, as it cuts both ways. When British people immigrated to Australia, the native Australians called them Pommies and hated them for moaning about Australia's not having certain British "virtues." British immigrants were accused of troublemaking activities of trade unions and expected Australia to be a Britain in the sun. It took some time for the Brits to realise that Australia is Australia and nothing else. However, the Brits are, at least, willing to adapt and learn. Pommy-bashing has gone down, I have heard. The reader may be confused and wonder which side I am on. I see two conflicting moralities at play. Only by recognising both can there be any chance of solving these problems.

Intolerance

I remember reading an article in a newspaper about the success of suburbia. It's like this: In suburbia, where people own their homes, people are free to live their own lifestyles and to paint and decorate their homes according to their own taste, whether that means having a Georgian front door with gnomes or statues in the front garden or something else. Everybody keeps themselves to themselves and generally minds their own business, but they know that they can rely on the people around in an emergency. In suburbia, people make sure that their children behave in a civil manner and are concerned with what the neighbours think, as the neighbours are a part of the community. At least this is the traditional ideal of English suburbia With this attitude at the forefront of people's minds, the suburban way of life works very well.

However, there was the case of Jack the lad joker, who put a modal shark's head through a roof. People in his area went along to the council and asked to have it removed. This is not a case of narrow-mindedness or intolerance but a case of enough is enough. "We are a tolerant, reasonable people, but you are destroying the very stable atmosphere that we wish to preserve." This is not a case of Jack the lad's being a threat, as people did not fear the shark's head. It's that the shark's head was not mounted simply as part of a house or garden design. It was a form of over-the-top antagonism (the term *antagonism* means here that Jack the lad was only trying to attract attention to himself, as in, "Look at me; I'm different," rather than expressing a genuine difference in taste) towards his neighbours.

When one sees people objecting to mass racial immigration, something similar in nature is involved, I believe. Not all immigrants are dangerous or evil, but they do tend to alter the stable atmosphere and successful lifestyle enjoyed by native residents up to that point. When I lived is South London many years ago, a family of Anglo-Indians lived up the road from my family. They dressed and behaved like the rest of the residents and experienced, from what I gathered, no trouble or abuse from anyone. However, more coloured people later moved into the area, which suggested a change in the area's nature over a period of time, as residents who remained there have told me. From an emotional point of view, people "perceived" coloured immigration, with the different cultures and fundamentalist religions it brings, as a shark's head through the roof.

I remember hearing an idealist say that we Brits should show more tolerance to coloured people and other minorities and behave towards them as if they were white. That has happened, to a degree, as there is now more emphasis on people's being more tolerant. But then I have heard that minorities do not want to be tolerated; they want to be accepted. Let's understand exactly what we mean by tolerance.

We all know that one cannot help what one feels, except under the most bizarre conditions. But one can help what one does. Tolerance makes the difference. It's like this: We all feel pain as a means of defence to warn the body when it is under attack. We would be the poorer without the ability to feel pain. When we have a tooth pulled and know that it's for our own good, this does not, in itself, remove the pain. The pain is there, whether we deem it useful or not. We need a temporary anaesthetic to have a tooth pulled. In just the same way, proving to people how nice immigrants are and how they can make the lives of the host country's natives peaches and cream is not going to remove people's feelings of antagonism towards immigration. If immigrants want to be accepted, then the way people feel must be changed. You cannot alter what people feel any more than you can turn a homosexual into a heterosexual. Tolerance is as far as many of us will go. To ask any more is to ask for to much.

One thing that bothers me about the concept of tolerance is that it can be confused with apathy. Apathy can be a dangerous thing in any stable civilisation. Impose tolerance on that which is intolerable and apathy will be the result. Apathy can rob a civilisation of its energy and its desire to grow better and stronger. *Beware of apathy.*

A Misunderstanding of Race and Politics

Jewish people are the first to say that racism can lead to a holocaust: Ban racism to avoid a holocaust. After the Russian socialist revolution, millions of people died of starvation and many others were sent to Gulag archipelagos. After the Chinese socialist revolution, many millions died after the social order underwent reformation. After the Cambodian socialist revolution, Pol Pot and the killing fields emerged. More people have been killed during socialist revolutions than died under Nazism. Does this mean that we should ban the welfare state because it might lead to the killing fields? I don't think so. The reason why we have holocausts is because dictatorship exists, not because of racism, socialism, or any other sort of ism.

Adaptation

There are many examples of adaptation. I remember the series of *Roots* on television. Although this may be fictional it does portray a philosophical point. One young black man, a new slave, was whipped after refusing to say the name conferred upon him by his owner. While he was being whipped, another black man plead with him to say "Toby," but he kept on resisting. Eventually, he adapted and said that his name was Toby in order to survive. Of course, the violins played and the trumpets blew as though a great victory had been won. However, the only victory was for the slave owner, not for Toby, for Toby had submitted to slavery.

Adapting in this way was not entirely in Toby's interest. He was victim of adaptation, not a winner in the face of it. When a species adapts to a change in the environment, it is not about the virtue of any individual. It is simply that any winner has no more choice over what its genes will be than the loser does. When early humankind was faced with cold weather, instead of becoming a slave to the weather and living and dying accordingly, human beings adapted by putting on clothes. They did not have to change themselves as a species; instead, they changed the "environment" by enclosing a warm environment inside the clothes. These human beings did not submit to the weather but kept it at bay. It's rather like an astronaut who benefits from an earthly environment inside a spacesuit or a diver that puts on a rubber suit to insulate himself against the cold water.

When politicians talk about adapting to the emergence of the Muslims and their religion, they wish us to adapt to Muslims' demands, as though this a great challenge for white British people. But this situation is not about putting on different clothes; it's about saying "Toby."

True adaptation is about isolating oneself against what one does not want, not changing oneself to suit the bullies of this world, but merely to be able to live free of that, that makes one feel pain or discomfort

To reiterate again what I already have said.

If a person does not like anything that surrounds them, "one gets over it" by keeping what one does not like away from oneself. One might believe, rightly or wrongly that one has a right to isolate oneself from unlikeable situations. Just as one would not have the right to expect someone to go into the cold without a coat, it would be wrong to deny the right of a person to wear a coat.

Chapter 12

What Is Natural, and Interbreeding

Natural Nature

I have become aware that many people of the left-wing persuasion tend to be obsessed with the word *natural*. I believe that there are two ways of looking at being natural. These follow:

1. Natural as in "red in tooth and claw" (of course, I do not say that natural is always good)
2. Natural as in "towards the best possible health"

Let's analyse "natural as in 'red in tooth and claw.'" Before human beings created civilisation, death was the natural rule of the day. Most babies died in infancy for many a reason. Before modern medicine appeared, even wealthy people expected to lose some of their children before they reached adulthood. In the wild, contracting diseases, having accidents while hunting, being killed by wild animals, being wounded and dying of gangrene, experiencing pain and suffering, including cancerous growths, and experiencing other horrendous occurrences are all *natural*. The fact that we might find herbs that will cure us is no thanks to the plants but an accident of biochemistry. Plants develop substances to defend themselves from insects or herbivores, not to do us a favour, not any more than a tree exists to supply us with wood so that we can make a suite of furniture.

Humankind has, over many thousands of years, perfected culture and himself to use all the materials around him that he believes (subconsciously) were put there for his benefit. I suppose one could say that the oil in the ground was put there by nature so we could use it to power cars and make nylon stockings. I suppose that the dung beetle could say that the great dung beetle in the sky (who made the dung beetle in his own perfect image) put the elephant on this earth to supply the dung beetle with dung. Well, what other reason could there be?

Let's now look at "Natural as in 'towards the best possible health.'" This means that certain behaviours and certain foods may be able to help organisms to develop the full potential of their genes. For instance, for malnourished children, extra vitamins may increase concentration, thereby enabling greater learning ability and, hopefully, more knowledge. Also a diet higher in protein may mean a healthier and faster growth rate and allow children to develop the ability to read and write – storing knowledge on paper rather than trying to remember everything they are told straight off the cuff. Using computers may release humankind from dull, monotonous, soul-destroying work, thereby allowing people to do more enriching creative work. Civilisation presents a very healthy way of life. If you doubt this, then try living with a primitive tribe that doesn't practise hygiene or have modern medicine. You will be astounded at the mortality rate. The reason that tribal people would say they are happy is because they do not know any different.

I remember seeing on a TV documentary a tribe of Indians who had got used to living in houses, living a more modern lifestyle. When an idealist (from the "Nature is best" brigade) said that it would be better and more natural for them to live in the jungle, an Indian replied, "You try living in the jungle. It's bloody hard living from hand to mouth."

It's often believed that the Amazonian Indians have all the answers to modern medicine. If they did, then why did they not out breed their food sources, and why did their children die of epidemics which occurred long before modern human beings emerged? No doubt, the Amazonian

Indians do have medicines that have some effect, but it was only when modern people came on the scene and analysed the chemistry that these medicines had any universal applications. I therefore say that, in this context, manufacturing vitamins, being able to read and write books, using computers, building research and development laboratories, etc., are perfectly natural things to do, as they will benefit humanity's future health.

Analysis of Nature

Over many years, people of the left wing in Great Britain patronised the native population, saying that we should have loads of hospitals to cure the sick and make the crippled walk so that they could live full and healthy lives, marry, and have children. In many cases, people get sick because their health is relatively weak. In many cases, this weakness is of a genetic origin. When such people are made well, the doctors interfere with nature. In the wild state, these people would die or be in such poor health that they would not be able to compete in life's struggles. They would end up dead or childless, or have fewer children than their brothers and sisters.

When you use modern medicine, you are saving people who would normally die and not reproduce. These people have children who inherit the same genetic weakness. They will be cured by way of modern medicine and then go on to have children with partners who may also have the genetic weakness. In many cases, this leads to human misery, as perfect cures rarely exist. Over a period of many generations, more and more people will require continuous medical backup to keep them alive and well. This will take its toll on the welfare state, putting a huge burden on healthy and unhealthy people alike. But remember, this came about as a result of using unnatural medicines. Humankind became unnatural when they discovered fire, wore clothes, killed from a distance, built homes from wood and brick, and domesticated animals for food, clothing, draft work, etc. If you believe in nature, then you can see that humankind is a very unnatural animal.

Of course, the modern Greens would say that making chemicals, motorcars, and aeroplanes is unnatural. But then, many years ago, people made alcohol and changed iron oxides into iron, which is part of chemistry. Also, horse-drawn carts and any metal objects could be called feats of engineering. These things could be said to be unnatural, especially if you happened to be a horse. In actual fact, if you analyse what the Greens define as natural, you will see that it indicates something that one can make oneself (or understand the process for its manufacture). If you cannot do these things, then the thing is not natural. The Greens would say that something is unnatural if it produces pollution and helps destroy the earth. Well, that is to do with overpopulation. Any human activity, even going to the lavatory, will cause pollution if the population is large enough.

Eugenics

We now live in a state absent of genetic selection, which is as unnatural as one can get. Our very existence is due to genetic selection, whether that selection has been natural or otherwise. If it weren't, then we would not exist in any form, not even as amoebas, since that organism was only produced by way of natural selection. Now, biochemists have come up with ways to scan human chromosomes for mistakes so that fewer unhealthy children will be born. And now, the "Nature is best" brigade is on the march, condemning these experiments as though the scientists were out to produce Frankensteins. The way we are going, without any genetic selection at all, we are more likely to produce Frankensteins galore as more and more genetic mistakes occur. The "Nature is best" brigade does not like scientists experimenting with human embryos, as they see this as interfering with a whole individual future baby. Well, I suppose they have a point. However, suppose the scientists were able to scan and select out the best egg and sperm before fertilisation, ensuring, as much as is possible, the best possible health for the individual. We as a species could practise genetic selection on ourselves just as Cro-Magnon may have done, but without the pain of there being any "losers." This would mean that, instead of our evolution having to fit passively with our environment, we could change ourselves as it suits us. I would suggest that this would not please Neanderthal man (or his modern equivalents) who worshipped nature far more vehemently than Cro-Magnon man did. And yet by doing this, we would be doing something natural, but without the pain and suffering of nature.

At least when humankind practises selection on human beings, it is not the irresponsible blind selection of nature. One never knows what those types of changes will lead to, including the "blind alleyways" that many an organism has been trapped in. As far as health is concerned, a great deal of conditions that the National Health Service (NHS) of Great Britain supports research for, such as short-sightedness, have a genetic origin. The NHS could help improve Britons' teeth by funding research in this area, since teeth that are resistant to decay may also have a genetic origin. Cystic fibrosis has a genetic factor. Indirect genetic weaknesses, such as a susceptibility to certain cancers, could also be explored. Even some forms of senile dementia could have a genetic factor.

Funding so much research drains the NHS of a great deal of resources, which could be used elsewhere if the above-mentioned problems could be solved by genetic selection. If we were to eliminate as many genetic weakness as possible, then this would leave the NHS free to treat accidental injuries and also invest in curing diseases like leprosy and malaria, which are still hard for even the genetically healthy to resist. I do believe that a resistance to leprosy is genetically based.

As far as selecting a child's appearance is concerned, the first question one might ask is, "Who would make these selections?" Another relevant question is, "What should the children look like?" I would say that it is up to the parents to make those choices. If parents wanted to have blue-eyed children, boys with dark hair, and girls with blonde hair, would that mean the fall of civilisation as we know it? I hardly think so. What other choices would we make when selecting the genetically based characteristics of our offspring? I think the best way to judge what is a bad genetic characteristic is to ask oneself, for instance, "Would I like to be born with [this] or [that] characteristic and go through life looking like [this] or [that]?" What adjustments would you, reader, make to your own body, if given the chance? Imagine for the moment a world full of people who are recognised as healthy and beautiful people. No one is undersized, weedy, un-athletic, disproportionate, or intellectually retarded. Surely, the world would be a happier place if people lived without inferiority complexes and the resulting problems that exist in our present world.

It is possible to have a genetic problem with the shape of the pelvis, which can affect the health of a foetus as well as the womb environment during pregnancy. The womb environment could be more closely monitored for too much of the male or female hormones, which can create homosexuality and mismatched gender identity. These two things cause a great deal of confusion and also wreck people's lives. I know it could be said that beautiful people can be conceited and "full of themselves," but this only happens because they are the exception rather than the norm. If everyone was born to exhibit nearly ideal health and beauty, then what would any one person have to be conceited about? It could be argued that people wouldn't appreciate health and beauty if everyone was healthy and beautiful. Would you choose to have weak and ugly people for this reason, allowing nature to run its course even though we could avoid it? Those who have eyes to see, ears to hear, and arms and legs to move can appreciate these things without having people around who lack them.

A reader of this book might ask, "What about the personality and character? Don't these qualities matter?" Yes, of course they do, but just because people are fit healthy and beautiful does not mean that they will lack personality or character. Since when did the weak and the ugly gain a monopoly on this virtue (although the weak and the ugly probably like to put this idea around)? If genes are to be found that will give people character, that is to say, control over oneself, then who would argue about manipulating those genes?

I must confess that while thinking about personality, where the balance of A, B, C, and D profiles is concerned, it sounds as if I'm engaging in genetic politics, given that how a person votes in an election may be affected by his or her profile balance. I must confess that I do feel a bit uneasy about this line of thinking. However, I believe that if everyone had an even mix of A and D, this would lead to a more stable society. Still, men would have a dominance of B and women a dominance of C, as occurs naturally. Genes that have the normal, harmless variations would enable a variation of individuals to keep society healthy The main thing to concentrate on is good all-round intelligence. But then, as I have always said, issues like genetic selection should always be discussed openly and pragmatically as time goes on, not practised as a state religion.

It is true to say that diversity in the human race is beneficial to future generations, especially given the new challenges humanity will face in space, but I would say that the people who are the most adaptable in such situations are fit and healthy, not the sort of people who have extreme physical deviations from the norm – those who often visit the hospital for medical care. It is possible for people to vary a great deal and still be considered beautiful, fit, and healthy by general consensus. Of course, one could say that Bruce Lee, Muhammad Ali, Daley Thompson, and many other exceptional athletes are (or were) unusual in build, but only because many other people are smaller and weaker in comparison.

Our hominid ancestors showed a desire and willingness to take risks. This is what drove the early apes to come down from the trees. The hominid could have remained an ape, safe, high, and dry in the canopy. Throughout human history, people have always taken risks. Up to now, these have paid off, and I believe they will continue to do so, so long as we monitor the situations and not take risks frivolously. Unfortunately, some people's attitude towards science is like the attitude of a religious fanatic, as though science were either a perfect god to be worshipped or a devil to be destroyed. Science, as I am sure the reader has heard before, is a tool that can be used responsibly or otherwise, depending on who's using it. As it is said, science is a good servant but a poor master. It is society's responsibility to ensure that science is in the right hands, not the responsibility of a few pundits who wish to take the place of the Almighty.

Choosing the Better of Three Evils

The First Evil

In the Third World, millions live in poverty. In India, the establishment spends millions on a space programme while many people remain poor. The British cannot afford our own space programme, given that we spend most of our wealth on national security and supporting the poor. To allow poverty in Great Britain or anywhere would be an evil.

The Second Evil

In order to prove a point about evil, I would say this: Suppose an airline decided to put pigskin-covered seats in its aeroplanes. This would deter Muslims from travelling on those planes. The outcry would be that the airlines are stopping Muslims from travelling on aeroplanes. My answer to that would be *no!* There is no law stopping Muslims from travelling on any aeroplanes. People may also say that Muslims are entitled to their religious beliefs and, therefore, introducing pigskin seats was discriminating against them. Well OK

It is also a belief of the middle classes in Britain to live within their means, which means that people have only as many children as they can comfortably afford to support. This is a practical belief, one held so as to give children a good upbringing, which, I would say, is more valid than a near superstitious belief about pigskins, which can do no harm whatsoever to a Muslim. Many of us know that if the middle classes are deprived of wealth, as happens in cases of over taxation, then they tend to have fewer children. Then it could be argued that to overtax the middle classes is to practise a form of genocide against them. It's all very well to say that people of the middle classes have a choice whether or not to have children. A Muslim has a choice whether or not to sit on pigskin. Therefore, I say that it is an evil to practise genocide against the middle classes by overtaxing them. This, I repeat, is an evil which the establishment practises at the moment.

The Third Evil

In order not avoid the first and second evil, Society could sterilise people if they have more children than they can afford. This would, of course, raise a great number of objections from amongst the left wing.

Analysis

The question here is which is the lesser of the three evils. Taking a short-term viewpoint, one might prefer the second evil. But looking at the problem with a long-term view, one may ask, as the middle classes are reduced in size, who is going to pay the tax in future if there are not enough middle-class people around to pay those taxes and subsidise the under classes? Remember the film *Idiocracy*.

Genetics and Racial Discrimination

It is an obvious fact that the greyhound race (breed) of dog is the fastest around a set racecourse. Suppose that a couple of greyhounds had a gene deficiency, which made them slower than the fastest dog of another breed. This deficiency would arise when both pairs of genes on both chromosomes were defective, which would mean that the gene would be recessive. In other words, the weak gene does not become apparent unless both weak genes are present. Suppose the gene deficiency was different in one greyhound compared to the other. Let's say that one had a gene deficiency affecting its breathing, while the other's genetic deficiency was to do with its heart's efficiency. Say that a poor man wanted to buy some dogs to start a breeding programme of his own in order to participate in the dog races. He had a choice of four dogs. Two of the dogs were the defective greyhounds, and the other choice was two mongrels that were faster than either of the first two greyhounds. Now, a person who judges organisms by their abilities would chose the two mongrels. However, if the poor man were a bigot, then he would prefer the two greyhounds, since these dogs have the status of speed. A geneticist would say that if the two greyhounds were to interbreed, then the young of the first generation would be reasonably healthy as their many great-grandparents were, but they would all carry two masked defective genes. In the second generation, one in sixteen greyhounds would be completely free of both defective genes, masked or not. These dogs would be as genetically healthy as others of the true breed and would certainly outrun any other breed. In this hypothetical situation, you can see how bigotry can work for the better.

Humankind evolved without the knowledge of science, so people had to make judgements without the knowledge that we modern people are lucky to possess. If people in the past made the right judgements by way of feelings that we were to call bigoted, then bigotry would survive. We have survived because our ancestors made the right choices, bigoted or otherwise. Therefore, one must always remember that some bigotry can be survival-enhancing knowledge. It would be true to say that sometimes bigotry works, so long as it's based upon deep instinctive feelings, not necessarily on values imposed by society.

Misusing ideas

Although I do not like religion, it is the mindless adherence of religion that I am against, not having an honest belief which coincides with a particular religion.

If a person said that she believed in having a national health service, I could ask, "How can you believe in a national health service? Don't you know that Joseph Stalin believed in a national health service and that his regime murdered millions of people?"

It is very easy to condemn an idea just because some cretin may well have thought of or practised it first, at some time in the past..

Humankind has been gifted, through millions of years of evolution, with a lot of little grey cells. It's about time people used them. Judge an idea as it stands and not by the Machiavellian arguments of the feeble-minded who reject anything that they don't really understand. Of course, it is possible to misuse an idea. I doubt that there has been any idea in history that has not been misused by some Machiavellian imbecile. Making a religion out of ideas or the condemnation of ideas can only lead to a waste of resources and the degeneration of humankind. If a person is rejected for a job because of his or her race, in favour of another individual of another race who is just as capable (which I am not recommending, by the way), it does not mean that civilisation will fall apart. However, rejecting the right person for the job in favour of an incompetent Machiavellian who flattered and brainwashed the interviewer by using "magic" words can very easily lead a company to disaster

Interbreeding

The Results of Interbreeding

Interbreeding between two races can be a blessing or a curse. When modern human beings interbred with Neanderthals, natural selection took the best of both breeds, I believe (and as I have already mentioned). But this was at the beginning. It is also possible to inherit the worst of both, depending on what natural forces of selection are involved and which individual produces the most offspring.

As I shall make clearer further on in this book, it is very easy for a Machiavellian, having been able to manipulate his way into a strong position in a group and thereby having a wife and many children, to claim himself a winner. It is all very well to be a winner as an individual, but the Machiavellian's survival and that of his descendants are dependent on the group and how productive that group can be.

If the group is saturated with Machiavellians, then who does the producing? (While most Machiavellians are capable of turning themselves to production if pushed, inventiveness and thinking up better ways of production is not their thing.) Also, a group has to compete with other groups, which may not be so stupid as to allow a Machiavellian to have his or her way. If the group does not survive so well, then surely the Machiavellian will suffer in the long term, as well. Try explaining this to a Machiavellian. He or she will tie you up in knots with emotionally twisted logic. Although the political left wing is in favour of the "great brotherhood of humankind" and would assure us that interbreeding will give future generations the best of both worlds, it could just as easily give us the worst of both. Only if you have survival of the *fittest*, the selective breeding human beings would have in the wild, do you *perhaps* produce the best of both worlds. Can you imagine people on the left or people who are politically correct agreeing to that? Well it would be "natural."

Is racial interbreeding natural? Idealists often mention that it is natural for different races to interbreed; otherwise, why would they do so? According to the biochemist, speciation takes about a million years to occur. That is to say that fertile young may be born from two populations that have been genetically separated from one another by no more than a million years. This is due to what is referred to as the biological clock. Mutations in the DNA occur over many thousands of years. These mutations build up until the egg and sperm of two separate animal populations will not create a fertile individual. This does not refer to the difference in structure of an animal, only to its DNA's chemical difference.

There are two species of frogs in the Amazon jungle that are structurally very similar, but the structure of their DNA is dissimilar. The difference in the DNA of these two frogs is far more different then the difference between the DNA of a whale and a bat. The two frogs have been separated by a greater time than the Whale and the Bat have been separated.

Two breeds of dog, such as a bulldog and a collie, can be more structurally different than can two different species of cat, such as the lion and tiger. Expert palaeontologists sometimes cannot tell the difference between the skulls of lions and tigers apart. Just because two populations can interbreed does not mean that they will be similar in structure or that it's natural for them to interbreed, especially if a genetically specialised structure is beneficial to the native population in a given area. The introduced population has no such genetically specialised structure. In that particular instance, it is natural and beneficial for the home team to reject the newcomers, although it would be natural for the newcomers to want to interbreed with the home team, since it is in their genetic interest to introduce the well-adapted genes of the home team into their own gene pool.

There has been some concern about an "unnatural" introduction of a species of duck into Great Britain. The new species is closely related to a native duck, close enough that the two could interbreed and produce fertile young. The introduced duck is considered a separate species because it would not interbreed with the native duck if people had not artificially introduced the species. I would ask the "Nature is best" brigade, "Are South Saharan Africans and Asians the same species as Europeans, given that they were only introduced into Great Britain by unnatural means, such as unnatural aeroplanes and unnatural passenger liners?" (I doubt whether any South Saharan Africans or Asians floated on logs to reach Great Britain). Would the "Nature is best" brigade say that every child in Britain who was of racially mixed parentage was the product of an "unnatural" mating? Returning to the subject of the ducks, I ask if separating these two breeds would be interfering with the natural brotherhood and sisterhood of ducks. The "Nature is best" brigade would say that since human beings evolved as a part of nature, all that they produce is natural for humankind to use, just as a bird's nest is natural for a bird to use. Well, then one could say that cars, trains, motorways, nuclear power stations, computers, and packaged food are natural for people to use, too. This kind of contradicts the original "Nature is best" philosophy.

It has been said that many a British soldier lost his virginity to a ewe in the field, much to a farmer's annoyance – and to the ewes', no doubt. I believe that some venereal diseases came from the Middle East, from people having sexual intercourse with *camels*. The mind boggles! There are paintings on cave walls from many thousands of years ago depicting men having sex with animals. I think that by now we should acknowledge that humankind's behaviour is not a simple case of natural versus unnatural. Let's investigate. When men have sex with ewes, camels, or blow-up dolls, they are not necessarily looking for permanent or meaningful relationships. On the whole they are only after sexual gratification or else want to explore what it would be like if . In judging men's sexual behaviour or sexual relationships as natural, one must look at the type of behaviour involved. Note that I am not comparing interracial intercourse with the above. All I am trying to do is counter the dogmatic belief that because one has a desire to do something sexual, this means that the behaviour is natural and, therefore, that one should do it. It reminds me of what Bob Hope said about homosexuality: "When I hear they are going to legalise it, I will get out of the country before they make it compulsory."

Where racial dogma is concerned, people express a great deal of self-righteousness. Does it happen that a white European man goes to a racially foreign country, marries a coloured woman, has children with her, and then, when the marriage fails, kidnaps his half-caste children and takes them back to Europe? I do not say that such a thing is impossible, but it is not the general behaviour of white European men to marry coloured women, let alone take their half-white children back to Europe, exhibiting the pride normally expressed by fathers.

However, it is not unknown for an Arab man to marry a white woman and then kidnap their children and take them back to the Arab country, bringing them up as Muslims and showing others, with a degree of pride, that his children are part white. Even Arabian kings and wealthy sheikhs are known to marry white women.

Speciation occurs only when similar organisms are genetically separated from one another for more than a million years. Speciation does not indicate the different structure of similar organisms. It is theoretically possible to sexually desire a member of another species if its structure is similar to one's own. This type of union is not perverse. It could equally be said that just because two races of the same species can have fertile young does not mean that the two are supposed to interbreed or that it would be beneficial for both to do so. It is simply a matter of who benefits. I would say, to sum it all up, that it is natural to do the best thing for one's genes but not for other people's. A man does not lose anything by impregnating a woman here and there, but a man does waste time and hard work in bringing up children whom he does not consider to be the best to pass his genes to future generations. Also, when it comes to choosing a mate, it is natural to go for the best quality one can get. People should accept that their quarry might not consider them to be the best quality. I consider it perverse to have children of a lesser quality than one could have when better quality is available. I would point out that what is considered to be the best quality, when all is said and done, is best left to personal opinion.

Individuals should decide for themselves. I would say that it is perverse for a woman who only wants a child, not a husband, to have a child with a man who is not, in her opinion, the best-quality man she could mate with (where others are available). Of course, I imagine that I will get a right Machiavellian dressing-down from the politically correct for saying these things, even though the politically correct know that what I say is true. The politically correct do not offer any facts, only self-righteous or moral indignation, and then they express, out of all proportion, the odd exception to the rule. *So, what's new?*

Chapter 13

Slavery, Censorship, Discrimination, and Baldness

Slavery

There does seem in the Western world a great guilt complex over slavery. Quite frankly, modern Western people did not invent slavery. A slave is generally defined as a captured person who is forced to work for no pay or reward. The first slaves, I suppose, were women. Many women today "humorously" would say that things haven't changed much in the last thousand years or more. The first male slaves were probably captured prisoners of war who would rather have been slaves than be killed. Of course there have been cases where the Romans would wage war for the purpose of getting slaves. It was generally understood that the enemy were expected to die, but the winning army did them a favour by allowing them to live. In Rome it would be possible for a slave to win his freedom by hard work, eventually.

Slavery is as old as civilisation itself. The pharaohs held slaves, as did the Mesopotamians, the Greeks, and the Romans, who enslaved the Britons (please note). Arabs once practised black slavery (and I have heard they still do, but this is hearsay). In fact, I do not know of any civilisation that did not, at some time in the past, hold slaves. Even "holier than thou" black men were known to have slaves. In fact, it was black men who sold black slaves to white people during the recent period of slavery. For that matter, black slavery in the American colonies would have been very difficult to manage if there were no black men from whom to buy black slaves. Can you imagine a few white men with *single-shot guns* going into a forest, taking a tribe's brothers and sisters, and expecting to come out alive? I think it is obvious that a tribe of black men would have been deterred from attacking armed white men just for the fun of it. But they would almost certainly attack after their territory was intruded upon. On top of that, they would not have allowed their relations to be taken as slaves. I believe that tribal black people would have certainly taken the risk of being killed in order to stop the white people by way of mass assault. In such a case, the white people would have considered this to be too risky. Black people were taken as slaves when one tribe surprise-attacked another, perhaps while most of the menfolk were out hunting. Therefore, the whites would have had superior numbers. Many black slaves died on the way to America after being deprived of food and water, but others committed suicide, thinking that it was better to be dead than live as a slave. The ones that chose slavery rather than death are the ones who left descendants in the United States.

In this day and age, the black population "milks" white people's guilt about past slavery, but are white people really responsible? Let me put it this way. Suppose a black man mugged me tonight on my way home. Should every black person feel guilty about it? Of course not. We do not take on the sins of other people just because they happen to be of the same race as ourselves. The black population could say that it was white people's ancestors who practised slavery and that, therefore, the sins of white people's forefathers are on white people's heads. But wait a minute. How many people in Britain, for instance, have ancestors who practised slavery? Most of the people who practised slavery lived in the colonies. Some of those men raped black women and had coloured children. A good proportion of the children born to black women slaves in the colonies may well have been half white. Therefore, I would suggest that the black population of Jamaica and the Americas probably have more ancestors who were slave owners than the average Briton does. Now, of course, a black person could say that it was Western society that enabled whites to practise slavery. Yes, that is true, but if you look back in time, then you will find that the average Briton had very little power over the establishment, which acted independently of public opinion. And anyway, how would the average Briton in the 18th century have known what was going on in the rest of the world at the time? Britons were doing their best to survive in a cruel world, one which was far harder and more oppressive for a working-class white man than it is for a black man living today in USA or Great Britain.

Censorship

Up until the early nineties, there were two forms of conversation about the Soviet Union. One was the public conversation, which went something like this: "We must all work harder for the revolution, comrade. We must all pull together. Are we not lucky to live under communism, unlike those poor people who are exploited in the West?" The private conversation went something like this: "Good God, when will it all end? The state demands more and more, and our living standards are not improving like they are in the West. We seem to be working only to stand still."

If Soviet citizens were caught saying the latter, they would be reprimanded, told that they were being disloyal to the state, and seen as being against the people. Such a thing could not happen in Great Britain, could it? I know that where race is concerned, two conversations do exist: a public one and a private one. The public one, when someone is in front of the media or an audience, goes something like this: "I am not against coloured people. I mean, they are all human beings like us, aren't they, so we must all learn to live together in peace and harmony." The private conversation is rather like this: "I don't so much hate the coloureds. It's just that I would like to get away from them so I can live amongst only white people. As for the "Pakis" and their bloody religion, I wish they would get out of the country and go woe and wail about bloody Allah elsewhere." When one censors anything like the speculative private conversation just mentioned, all one does is hide grievances, which fester undercover. One does not destroy them. When you allow people to express what they feel inside and you respect their feelings, you are in a position to reason with others and come to some workable solution.

Discrimination

Discrimination has been with us as long as human beings have existed. Most philosophers would agree that, strictly speaking, we can never know anything and that all human experience with the world is subjective. All is measured by approximation. We recognise friends and relatives by judging their appearance. This is extremely sophisticated. Unlike computers, which ascribe this or that value to all things, human beings judge everything by way of approximation. Human beings cannot operate without judging everything in every aspect of life. We have evolved to judge. Also, every animal judges. If an animal or person wishes to jump over a gap, then it does not (to let the cat out of the bag) measure the gap with a tape measure. It makes a judgement by calling to mind past experiences of a similar situation. Many an animal has died by making a wrong decision (judgement). We have all evolved to make decisions by discerning the best possible bet, leaving as little to chance as possible. I repeat: We make the best possible bet.

When choosing a spouse, lover, or friend, and in business and when buying goods, we can never know anything for sure, so we go by our own past experience, others' reputation, and what we have heard others say in order to determine the safest bet. We have to be careful about what we hear since it is very easy to be led astray by misinformation. Beware of the statement "There is no evidence for this conclusion." Evidence depends on the extent to which one demands proof, since absolute proof is absent when dealing with the social sciences. It is very convenient to demand absolute proof, pure scientific measurement, when one is not dealing with a pure science. However, if people had always held to that standard, then medicine and the social sciences would still be in the Dark Ages. Also, it is very convenient for those in power to demand proof from the general population when one of us expresses an opinion. The people in power are the only ones who have the resources to get proof, but they refrain from even trying to do so, for political reasons. Proof will rarely emerge if one does not look for it. One could say there is no proof that UFOs do not exist. All right, but there is no proof that they do exist, either. In such a situation, one can only use one's common sense, which may be a somewhat alien concept to pundits. It does seem unlikely that UFOs exist. Therefore, in the usual course of day-to-day living, some people live with this assumption.

Often, when someone is interviewing for a job, he or she faces an interviewer who is ignorant of his or her quality, looking instead for official qualifications. But official qualifications do not alert the interviewer to qualities such as intelligence. Many of people's qualifications, e.g. academic degrees and professional experience, were gained by memorising facts and studying hard. These things are not intelligence. Also, people who are willing to accept whatever a teacher is willing to put before them are normally lacking in individual imagination and are less likely to experiment with new conditions. A candidate's character should be considered during job interviews. A person with impressive qualifications may be a drug addict or alcoholic, for example. Sociability is another desirable characteristic for a job candidate. The job may entail communicating and working with people on a rather delicate level. In addition, job candidates should be assessed for reliability. This may include good physical and mental health, good timekeeping abilities, and the skills to do the job required.

As interviewers become more astute about choosing the right candidate, it is also possible for interviewees to acquire more tricks of the trade and then use them to fool the interviewer. This is where Machiavellianism becomes a *dangerous* factor for the interviewer, who can never know whether an interviewee's behaviour is natural or designed. Say that you are an interviewer and have found that 90 per cent of people with blue eyes are gifted with a desirable quality, and that only 10 per cent of people with brown eyes have the same desired quality. Faced with two people with equal qualifications and no way to test the desired quality there and then, you almost certainly discriminate in favour of a person with blue eyes. (Of course, you would then have to watch out for people wearing coloured contact lenses.) This is not a case of bigotry. It is simply playing the odds.

An instance of discrimination found in nature is when a young mammal or bird catches a stinging insect. The body of a stinging insect is normally patterned. For example, bees and wasps are striped. A mammal or bird associates the pain of a sting with the pattern on the insect's body. Therefore, the creature learns to leave insects with striped bodies alone. This means that it will also leave alone such insects as the harmless striped hover fly, thereby missing out on a would-be meal. Of course, it's worthwhile to miss out an occasional good meal in order to avoid being stung. If this were not the case, then animals would not have evolved in the way that they have. If a creature were able to discriminate between hover flies and stinging insects, then all the better, but doing so may use up time and energy, which the active organism may not have to spare. Why should things be any different for humans? The real problem with discrimination is that people do not have enough information to discriminate thoroughly and, then to make a correct decision. In a world where some people fear Big Brother, how can one get all the information that one requires? Furthermore, even if an interviewer had all the information about every candidate for the job, would he or she know enough to be able to tell who would be best for the job – or what the discriminating criteria should be?

Baldness

The writer of this book is cursed with a bald head. I do not say that bald is beautiful or that I am proud to be bald. On the other hand, I am not ashamed of being bald since my baldness is not my, or anybody else's, fault. I do not accuse "hairy tops" of being jealous of my baldness when they get a haircut. It seems contradictory when people say how good baldness is but then turn po-faced when ridiculed by a "baldist." At this point, the bald person may say, "I cannot help it that I am bald. It is not my fault." I am proud to be British. When I am called a Pommy, Limey, or Brit, I don't take offence, even if the person wished to give it. Why should one take offence at that of which one is proud? It has been found that people with bald heads are considered less successful, socially and business-wise, than others, even though this is only evidence of a subconscious feeling. Bosses may discriminate against bald people and favour people with full heads of hair. I am not going to go out with a placard and complain about it. People are the way they are. I myself may even subconsciously discriminate against baldness. Who knows? I would also point out that to make fun of racial characteristics is considered to be in bad taste. Lefties are often ready to attack anyone who ridicules people of a different race. However, people who have bald heads are open targets and have no defence whatsoever! I suppose that we baldies just have to live with it.

With all the talk about black being beautiful, and with black people's pride in being black, some black people easily take offence at the merest hint of racism. The reason why I do not call black people niggers, wogs, coons, etc., is because I do not like to be called a baldy. As some bald-headed people are not proud of their baldness, black people, I believe, are not proud of being black (although I am not suggesting that they are, or should be, ashamed of it).

Chapter 14

Women's Liberation

As far as women's liberation is concerned, I do sincerely sympathise with women for having suffered many injustices at the hands of men. I am all for protecting anyone against brutality, sexual harassment, rape, inequality, discrimination in the workplace, and the like. I am, however, a bit dumbfounded about the attitudes of women's libbers where marriage is concerned. The idea that men should do equal work in the house and look after children is somewhat perplexing, especially when it comes to breast feeding. Governments are not going to order men to look after their babies while their wives are at work. The whole idea seems ridiculous. How can a state legislate what people do in marriage? Surely, this is something to be determined between a couple before they marry.

From what I gather about many women, the reality of what they want in the long term contradicts their fantasy about what they believe they want. As I think I made clear in the chapter about psychology, there is masculinity in women as well as femininity in men. Women tend to judge the value of work and career in light of masculine values; therefore, they see housework and bringing up children as low-grade and unimportant. But this is only a masculine value, not a *true* value (if one exists). Married women who do not have a career tend to label themselves as only housewives. Of course, some women are happy with this role. But this "only a housewife" attitude is a man's attitude, not an *objective fact.* Even so, since some women judge work by men's standards, women's self-esteem is not helped. I am not suggesting that, as a man, I feel that child-rearing is the most important job for a human being. I would find the job quite boring. But I must say in all objectivity, and I do not wish to sound patronising, that raising a human being from birth to adulthood is the most important, difficult, and demanding job anyone can do (unfortunately, this does sound patronising. *Sorry*).

Many women want an equal opportunity to be promoted in workplace and receive the same wage packet that their male counterparts receive. For every woman who is promoted, a man is not. The men passed over for the job, therefore, work at a lower level than the promoted woman. I remember hearing women in the United States complain that there was a shortage of eligible men for them to marry (what they meant, I believe, is that there was a shortage of single men who earned comparably high salaries). One can't blame the well-paid eligible man for marrying a woman who has no career, because, in such a relationship, there is none of the trouble that comes with having conflicting careers.

What the hell do women's libbers expect society to do about this problem? If women's libbers were to say that men should stay at home and be house husbands, then would they seek changes in the law to force men to work only in the home? I hardly think so. Even if underpaid men were willing to stay at home, I would suggest that women, being what they are, would grow tired of a man who talked about the baby and how his latest recipes turned out. These women would have an affair with a right chauvinist pig. So, what's in it for a man who is willing to be a house husband? Can you imagine a woman working next to men who hold high-power positions, make important decisions, and are in control like she is, and then going home to find her husband holding the baby? "There will always be exceptions in all things".

One option is for a woman to have a career and also bring up children and look after the house by herself. Nice idea, if it works. However, such experiments sometimes result in a woman's hiring someone else to look after the kids and do the housework while she is at work. This means that the businesswoman cannot pay her helper a salary equal to her own. The hired hand will earn a lot less. If a woman is supposed to be able to choose between having a career and being married, then only a woman who needs money badly will be willing to look after someone else's kids. If a woman likes children very much and does not want a career of her own, then she would prefer to have kids of her own and look after them, not somebody else's children. Many career women hire poor domestics (poor when compared to the career women). This only exaggerates the differences between the rich and the poor – a return to the conditions present in *Upstairs, Downstairs*. Women simply cannot win – well, those who expect to have a marriage and career can't win. (There will always be the odd exception.) The principles stated in this paragraph apply to women who are ambitious and reasonably well-off and who can afford to be bolshie about their independence from men and the state. But a great deal of women are single mothers who are poor and dependent upon the state. What this means for the future I will not comment on, since I am not aware of all the facts, e.g. whether these women are single by choice or whether their children's father abandoned them. It is very easy for a woman to say that she is on her own by choice, as a matter of face-saving and pride, rather than admit that she was abandoned (or rejected). What this means for future generations of children, only time will tell.

It has come to my notice that in contemporary plays and films, career-women characters tend to treat their husbands with indifference, if not with disdain or contempt. The female characters at least seem less loving than traditional housewives in older films and plays. While this is only a matter of what's written in the script, I do wonder if this is a reflection of the way things are? Is there is an element of truth in this?. Are writers, reflecting the general trend that women must see men as being in a superior position to themselves before they respect them. I suppose that there are always exceptions. I have observed that when a young woman is ask, what she looks forward to in life, she tends to say that she wants a career and only a career. If a young man is asked the same question, he tends to say that he would like to be a doctor, architect, engineer, lawyer, and so forth. In other words, a young man is willing to be specific about what he wishes to *do*. It appears that young women are more concerned with the idea of a career, as it denotes status or provides one with the identity of being a career woman, rather than with the desire to participate in a specific activity. A man may enjoy the status that comes with his chosen career, but he does not feel the need to have a career for the sole purpose of achieving a certain status. The strange thing is that no matter how much women say they are like men, their desire to be like men shows their real feminine difference, which only confirms the difference between feminine and masculine attitudes.

There is a gender gap in *performance* in schools nowadays. Girls *outperform* boys. I put the word *performance* in italics for a reason. I have often noticed that the use of certain words reveals something about the user of those words. The word *perform* is used to indicate actions that are practised in public. One may perform on stage or on an athletic field – in front of an audience. The idea of *performing*, in this case, is to show off to the audience. Does this reveal something about the nature of women teachers who use the word *perform* in the above context? It is only natural for women to be actresses. In Western civilisations, some women dress for attention, whereas some men tend to dress rather formally and unobtrusively. What does a woman hate more than being abused? Answer: being ignored.

I would strongly suggest that the current trend of pushing girls forward and into future careers is more a case of encouraging them to *show* what they can achieve rather than encouraging them to make breakthroughs in science, engineering, or any other field. It's as if women's lib is more concerned with hyping female superiority and humiliating the male ego than it is with human progress.

Men earn the most degrees from universities that carry out genuine research. I am not suggesting that women are incapable of getting degrees. Far from it. But on the whole, the masculine side of a human being is responsible for the natural desire to pursue scientific and engineering-based investigation.

I say the following as a warning about the future of humankind. Considering that the women who have the personal desire to pursue careers have fewer children than those who prefer to remain as housewives, women who have ambitious personalities will be out bred by their stay-at-home-mother contemporaries. Will ambitious women die out, in evolutionary terms? If this were to happen, then no harm would come to the human gene pool. However, if women continue to choose careers because their intelligence is higher than other women's, and if things carry on the way they have been doing, then fewer babies will be born to the most intelligent women. This will have a demographic effect on the average IQ of future children, affecting girl children as equally as boy children.

As far as teenage girls outdoing boys at school is concerned, I would like to point out that chimps are out and about socialising when human babies of the same age remain helpless and somewhat gar-gar. Does this mean that chimps are superior to humans? Well, no, given that a human brain takes a longer time to develop properly. A boy's brain does not develop properly until he is in his early twenties, whereas a girl's brain develops in her mid teens.

In the past, a certain type of woman tended to hold a decent man in contempt, only having respect for the bully or the extreme chauvinist. I believe that some women are like this because their fathers were rather old-fashioned and brutal. A mother would defend her husband by saying, "He is a man," meaning that that's how men are. This would give a girl the impression that men who do not behave brutally are not real men but wimps. This type of girl, once she became a woman, would not marry a "wimp" but would eventually marry a man who was like her father. She might eventually divorce him once she came to realise that she would be better off single. Rather than admit to marrying the wrong man, this type of woman may excuse her divorce by saying that all men are chauvinist pigs. Even if the most reasonable man happened to say something to the effect that men and women were different, she would instantly write him off as just another chauvinist pig if she hadn't already written him off as a wimp.

I remember reading many years ago about a survey that sought to find what men and women do best. The results indicated that women tend to make the best administrators since they tend to work more tidily and are better at communicating than men. The survey found that men tend to do better in research and development and production. This is, of course, a generalisation. It has been found that when a woman takes on an executive role within a company and aims to build the company up without trying to play at internal and external power games, the company does, in general, do well, perhaps better than a company directed by a status-seeking male (or a status-seeking female, for that matter).

I would add that it is necessary to take a calculated risk now and then, but it must be calculated. Without taking any risk at all, a firm can become sterile and bankrupt, leaving its people's ideas unrealised. Another point I'd like to make about women is that has traditionally been a woman's job to give assistance to a man, filling a role such as secretary, nurse, and so forth, and generally clear up behind him. Espousing the idea that women evolved genetically to fill this role would tend to outrage women's libbers. However, I cannot help but notice that women seem compelled to complain about a man's untidiness and working in a muddle, even though women themselves are not affected by the mess. They seem compelled to tidy up after men and, at the same time, complain about having to do the work, even though no one is forcing them to do it. If women think that that they are the same as men, then why do they bother about men's messes? Let men clear up their own mess. No one is forcing women to do it.

I have noticed in the press that people express outrage when a woman of over fifty years of age has a baby. The general complaint is that the mother may be incapable of looking after the child or that she may die before the child is mature. Good point. But this sort of thing often happens in the Third World. How many mothers in the Third World can guarantee being able to feed their own children? I would like to point out that I would rather be born to a reasonably well-off fifty-something woman living in a Western country, who could make provisions in case the worst should happen, than to any woman in the Third World. In the Western world, if one is left an orphan, one does not necessarily end up starving, illiterate, ignorant, and, perhaps, dead. If you wish to make rules saying that people should not have children unless they can feed, look after, and cope with the continuous development of the child, then I suggest that you start with the Third World – and impoverished parts of the Western world, too. Many young parents in the Western world are too incompetent to bring up children.

As an aside, I would say to those women who are dissatisfied with the men of today is that

(The following is a tongue in cheek comment, although there is an element of truth in it.)

> *Women breed the type of men that they choose.*
> *Blame your grandmothers for the men of today,*
> *and blame yourselves for the men of tomorrow.*

In Praise of Women

I believe that marketing is first concerned with wish fulfilment, creating a demand in the marketplace for something that people are not aware of until it is offered. Perhaps the demand has always been there but the industry hadn't been aware of it. I believe that this is a case of an unexploited feminine virtue, especially where traditional feminine needs for the home are concerned. My own mother has complained for years that things in the kitchen and nursery seem to have been designed by men, not women. I remember seeing a new type of nappy for older infants advertised on the TV. The nappy was for older infants going through a period of being weaned of nappies. This type of nappy is made similar to a pair of pants and is worn in case of an accident. When my mother heard about this, she pointed out that when I and my brother and sister were of that age, she ran up a pair of padded pants on the sewing machine for this very reason. It would appear that today's marketing people are about fifty years behind not only my own mother but also, perhaps, many others. I believe that kitchen and nursery supplies have become so advanced because more women now influence those industries. This has helped not only women but also single men and other people who live on their own. Need I say more?

I honestly believe that women are needed in industry and the commercial world. At the same time, this does not mean that it is bad for women to be housewives at some period in their lives. I would say that when a girl leaves school, it is good for her and society alike if she goes out to work. But when she marries and has children, it is best (for both her and her young children) for her to be a housewife until her children are old enough to look after themselves. At that point, there is plenty of time remaining for a mother to return to work with a good number of skills and experiences in hand for industrial and commercial pursuits. I do not wish to sound patronising when I say that if I were in charge of a company, I would hire middle-aged women for administrators, as they tend to have a greater sense of order than men and also tend to be down to earth, using practical and common sense. I know that when I have been asked a question about how something can be done, I tend to think in the professional sense of a long-term solution rather than of a one-off, here-and-now solution, the former of which women generally present.

Chapter 15

Sexual Perversions, and Immigration

Sexual Perversions

I will always question people when they say things which do not add up. It seems to me that women are the first to point the finger at men and cry, "Pervert." *Pervert* originally meant a person who preferred to engage in a type of sexual activity other than full sexual intercourse with a member of the opposite sex. A woman who prefers some sexual activity other than full sexual intercourse with a man is also a pervert, then. Let's suppose a man and woman are alone together. The man is enjoying looking at the woman and asks to make love. If the woman refuses, then the man is not able to enjoy full sexual intercourse with her. Question: Is he a pervert? Obviously not, since he is only too willing to have sexual intercourse with her. Now, suppose that the woman said that the man could do what he liked with her so long as he paid her a sum of money. Say that it turned out that paying her would cripple his finances for some time to come. Would he then be a pervert if he said no? I would say no, since his desire to have sex is not diminished; it is only the price that presents a problem. If the man had to pay the same amount of money to look at the woman as he would have paid to have sexual intercourse with her, could he then be called a pervert if he chose only to look?

The quintessential part of the argument is that a pervert is a person who, when given free choice of the two options, prefers to engage in sexual activity that is different from sexual intercourse with someone of the opposite gender. I would also say that a man who is ugly or unattractive and who is unable to seduce a woman into bed is not perverted. I would ask any woman who points her finger at the dirty Mac brigade, who may enjoy ogling women, "Why don't you offer your body to them for full sexual intercourse? If you are right in believing them to be perverts, then they will refuse and you will prove your point, thereby having nothing to worry about." I do not believe that any woman would take the chance. Women know, deep down inside, that a good proportion of the dirty Mac brigade are not perverts but are only losers at the sex game.

There is a tendency for women to prefer the courtship aspect of sexuality, that is to say, a man's spending money, time, and attention on a woman rather than immediately trying to have sexual intercourse with her. It has been said that many women, after a night out with a man, will go to bed with the man because they think they should do so as a way of paying him back for the evening out. This reason supersedes any enjoyment she may find in sexual intercourse. If women enjoy the courtship of love rather than sexual intercourse, then is it not they, not men, who are perverts? Is it not women who turn men into the *apparent* perverts they claim to despise, who are perverts? It's all very well to say that if men want sex, they should court women harder. But even then, women only give men a possibility of sex, not a guarantee of sex, after men spend their hard-earned money in courting women. It is, as I have described before, a normal Machiavellian trick to call people dirty names and to degrade, humiliate, and undervalue people so as to manipulate them for one's own selfish and, perhaps, "perverted" ends. I would say to women, Before pointing your finger at the dirty Mac brigade, think about what I have said here and look at yourself in the mirror. Remember, according to the modern feminist, men and women are all the same. Therefore, women are supposed to desire sex as much as men, aren't they?

Immigration

Why do people immigrate? Why do people emigrate? Some white people native to Western countries wonder why so many people from Third World counties wish to immigrate to relatively cold Western countries and also wonder whether they should have the right to do so. White people immigrated to Third World countries years ago, so what is so different about Third World people immigrating to First World countries? When white people immigrated to countries abroad, they were not attracted by the societies of those countries. If any Third World county had been totally uninhabited, it would not have made any difference to Western people's emigration, although, in the long term, it would have caused less trouble. White people did not explore Africa, the Americas, and Australia with the idea in mind of living in peace and harmony with the indigenous population and hoping to become part of it. Also, the ideas that white people picked up from people of Third World countries had only to do with cooking recipes, how to use a variety of different vegetables, and information about the geographical nature of the land. These countries changed very little until white people first arrived, apart from the fact that a number of animal species were killed off.

However, would a Third World person, if he could have, have immigrated to Britain, for example, when it was still covered with oak forests? If Britain were without shelter, civilisation, and the trappings of a welfare state, how many third world people would still immigrate to Great Britain? Perhaps a few might, but only a few.

People migrated because they sought that which Westerners created. Western societies offer a freethinking democratic system, which was created by people who did not think that a dogmatic religion was the be-all and end-all of human society. Immigrants to Great Britain want to practise the very thing (namely, religion) that made their own countries of origin an intolerable mess. People with the immigrants' mentality created those intolerable messes in the first place. (This is very similar in principle to the townie who moves to the country, as will be explained further on.) Therefore, to nurture the idea that immigrants' ills are due to the "wicked" white population, that is to say, if the white man disappeared completely or did not exist in the first place, then the ethnic population would live the life of Riley, is no more than to delude oneself.

In 1946, after the Nazis (who were exaggerated ego) were defeated, the victors experienced a euphoric feeling of togetherness (a kind of exaggerated self), where thoughts of the brotherhood and sisterhood of humankind abounded. Amidst this feeling of euphoria, laws governing immigration and nationality were laid out so that every person who lived within the territory of the British Empire had universal citizenship. Unfortunately, decisions made amidst the euphoric joy of the moment can be somewhat unrealistic in the long term and in the cold light of day. Britain's extending universal citizenship was not meant to allow huge, uncontrolled waves of immigrants into Great Britain. It was simply a law that provided a sense of universal dignity, nothing more.

Some older West Indians claim that in the late fifties and early sixties, the British government put posters up in Jamaica inviting Jamaicans to Britain so that they would come to work and help build up the economy. It was claimed at the time that Britain was experiencing a shortage of labourers. When Jamaicans arrived in Britain, they unfortunately found themselves unwelcome. (Encouraging immigration to Great Britain was the British government's idea. Although British politicians are elected by the British people, immigration was not an electoral issue at the time.)

The original idea for inviting Jamaicans to Great Britain assumed that they would stay temporarily, earning money for themselves and their families so they could return with new skills and money in their pockets to make a better life for themselves in their own land. (Just because a landlady in a seaside resort advertises for people to stay at her boarding house does not mean that the tenants are supposed to put down roots in her house forever.)

To say that Great Britain needed more labourers at the time is utter nonsense. The more people there are in a country for greater production, the more people there will be to consume what is produced. The only justification for immigration is if the prospective host country is underpopulated, which is hardly likely for a country like Britain. If the powers that be at the time had resorted to automation in some form or another instead of for the cheap labour of immigrants, then the British economy would not be in the mess that it has been from time to time. One question that people ask in an attempt to justify immigration into this country is, "What would happen if all the immigrants who came after World War II – and their descendants – decided to leave Britain tomorrow?" Well there's an objective question.

I would reply, "The same thing that would happen if they all decided to come here yesterday."

Someone might reply, "But they didn't all come here yesterday."

I would answer, "And they are not all leaving Britain tomorrow, either." Just as a society adapts to immigration, it can adapt to emigration. It would be true to say that the British society has not only survived for thousands of years but has also *progressed* without <u>MASS</u> immigration. I hardly think that it would fall into disrepair today had it not been for immigration over the last sixty years or so. There has been immigration to this country in the past of course. First you had the Celt, then the Romans, then the Germanic tribes, Anglo Saxons, Danes, Vikings etc. These people created Great Britain.

The pundits point out that there has been immigration from France (Huguenots) & Dutch, which came here with skills that were unknown in this country, which increased the wealth & knowledge of this country.

However these people came from the same place that our ancestors came from, they could therefore be called not immigrants but **"latecomers"**.

Someone arguing against me may point out that many immigrants fill roles as doctors and nurses. I would point out here that it does sound marvellous that ethnic doctors and nurses came to Britain to sacrifice themselves for the British people, but they were probably needed more in their countries of origin. Or should I say that people in the First World are only making things worse for people in the Third World by allowing many people who are needed in their own countries to immigrate – and that their so-called sacrifice is nothing more than selfish financial interest, like that of those British doctors and nurses who emigrated to the United States?

If Great Britain were to use financial incentives to entice British emigrants who are doctors and nurses to return or, more likely, to stop British doctors and nurses from emigrating in the future, then the problem would be easily solved. (It's quite possible that the well-trained doctors and nurses who emigrated to the United States helped to prop up the US private health-care system. If the United States had to train all its own doctors and nurses, then it might not be able to boast of such a "perfect" health-care system. It does seem perfect at the moment for one who is rich enough.)

Now, of course, it could be pointed out that if Great Britain paid more to doctors and nurses, then people would have to pay more in national health charges. Yes, of course, but on the other hand, if Great Britain did not have so many immigrants, then there would be cheaper housing available for all and, therefore, less to pay in mortgages. With fewer people in Britain, the cost of land and housing would be lower. Also, there would be fewer cars on the road and, therefore, less of a need to build more roads. I think that this would more than offset the cost of extra NHS charges and any other extra cost, don't you? An opponent may lecture me by saying that the coloured population is in Great Britain to stay and that wishing them away is not going to alter anything, so we Brits must all face up to the fact and learn to live with reality. Well, unfortunately, passive racism is here to stay, too, and wishing it away won't make it go away. Just as white people in Great

Britain are expected to tolerate people of other ethnicities, people of other ethnicities will have to tolerate a certain amount of passive racism in British white people.

I am not suggesting that people should accept race-based violence, only that people cannot help what they feel. If the white population, as a whole, has feelings of racism and withdraws into itself, refusing to socially integrate, then that will be a reality the coloured population will have to face, also. If, of course, the reader believes that the different racial populations of Britain are living together in peace and harmony, then he or she may ignore my writing and have absolutely nothing to worry about and, therefore, nothing to argue about.

The Townie in the Country

It's an old story with which many are familiar. The townie visits the countryside, getting away from the city and all its problems (high-rise flats, overpopulation, noise, hustle and bustle). "Oh," the townie says, "wouldn't it be nice to live in the country and experience the peace and quiet, the clean air, the green fields, and the little birds twittering in the trees?" When a townie moves to the country, he or she naturally finds themselves happy at first. But then they find that country life is rather Spartan. "Well," the townie says, "I am civilised, so I naturally expect some civilised facilities." As more and more "civilised" townies move to the country, they ask for better roads, more shops, proper sewerage, electricity, plenty of parking in the village or country town, and so forth. They end up ruining exactly what attracted them to the country in the first place. Also, they bring their vices with them. I am not suggesting that I am any different. As a Western man, I am only too aware of my own weakness.

A Third World Person in the Western World

In the Third World, before "wicked" white people came to colonise, natives lived a primitive lifestyle. The majority of their children died before reaching adulthood. However, "wicked" white people gave them "unnatural" medicines and facilities, which enabled them to grow and prosper. However, their "natural" desire to have large families was not diminished.

In the Western world with modern "unnatural" medicines, people began to overpopulate before realising that they had to exercise self-control. "Unnatural" birth control is the norm in Western societies. In countries like Brazil, peasants who live on the edge of starvation want more land, so they move into the jungle to find for more land to grow food. "Naturally," Brazilians have their larger than average families, and when the sons of the family come to maturity wants a little bit of land for his larger than average family.

(Yes, it is true that westerners have cut down trees and spread concrete over large parts of the land, but the land is still fertile. Western people have been forced to limit food production after overproducing crops in an overpopulated land.) Yes, the Western world has plenty of land because its people did the "unnatural" thing and controlled the birth rate after considering the technology at society's disposal and after looking to the future and realising that the resources were not in infinite supply. Someone reading this may say that westerners are responsible for messing up our own lands. Yes, but westerners were the first to industrialise. People in Great Britain built an industrial society from a low industrial base, which is far harder, because, in this case, people were building blindly, not knowing how things would turn out. Wood was first used for fuel until Great Britain had the technology to mine coal and effectively turn it into coke. The Third World has the advantage of learning from the Western people's mistakes. That is, they have Western experience to fall back on. If the Western world had not industrialised, then the Third World would not have benefited from the knowledge it has received from the West. When Third World people immigrate to the West, many of them go on to have large families, which only survive (if they are not financially successful) with the assistance of the welfare state. However, people living in a western civilizations are only able afford to have large families, supported in part, by the welfare system, because others do not have large families. Although this may be a gross generalisation. It is the generalisation of a society that makes a society what it is.

There has been a growing awareness of right-wing nationalism in Europe of late. The general view is that liberal-minded people like myself love Britain for its people's tolerance, self-control, and understanding of various points of view. But these are not the characteristics of "extreme" religious people. You will find that such religious people, whatever their religious persuasion, sink themselves into an abyss of wallowing devotion.

It is quite obvious that although they are bound to be influenced by Western society, they stick to anachronistic beliefs. The trouble with religious tolerance is that by tolerating people who are not by nature tolerant, a tolerant society changes into an intolerant one, as different types of people have trouble tolerating one another. What Muslim says that a Hindu is just as religiously correct as a Muslim? What Hindu says that a Muslim is as religiously correct as a Hindu? Of course, many religious people will appear tolerant while they live in a liberal society that protects them from people of other religions, but beware the time when liberalisation falls from grace. Then, whatever religion holds power will assert itself as God's only religion. I would say that no organised religion is religiously correct since its followers eventually impose their religion upon others, when given the chance (or the power). I would say, that true liberal-minded individuals hold a belief in the freethinking, free-feeling person. People of any religion that happens to be composed of a minority group, talk about universal tolerance and respect for others' religions, but as soon as a religion becomes all-powerful, the less its adherents respect any other religion. Unless the other religion is so small that they hold no threat to the establishment.

During the Cold War between East and West, the Third World was in the middle of two powers and was subsequently fought over. The fight was for the hearts and minds of people of the Third World, in case they should side with the opposition. People in the East and West alike fell over themselves, trying to prove how non-racist they were. This meant that it was governmental policy not to appear racist in any way. This created a people obsessed with non-racism. Even after the Cold War ended, the obsession remained. Great Britain has allowed immigration to occur to the point of insanity. A country can only accommodate a finite number of people. When the time comes when Great Britain can no longer expand in population, people born and bred in Great Britain will not be able to have as many children as they would have had if there had been no immigration. Therefore, people who would have otherwise been born and lived here will not exist because of immigration. This means that people who worked, fought, and died to make Great Britain the way it is will not leave as many descendants behind as they would otherwise have done.

I will likely be accused of being a bigot for believing that people who have children have a right to reproduce their own genetic kind. Why are men so concerned with paternity? Why should it matter if the child is the milkman's?. Or why would couples be concerned about test tube babies being mixed up in the laboratory. A baby is a baby isn't it?. I hope I have made my point about how our high and mighty political ideals do not match the day-to-day reality.

Chapter 16

Machiavellianism: The Continuous Threat and the Economics

I would point out that my contempt for Machiavellianism is only directed at those who use it for their own selfish purposes. Such a thing does civilisation no good. There are Machiavellians whose nature it is to sincerely help people, give them confidence, or inspire them to do better things without seeking profit. It is the latter type of Machiavellian to whom I apologise for any offence perceived in this book. The evil of Machiavellianism entails not only the out-and-out political lies but also the words used for the sole purpose of distorting the truth in favour of one's self-interest. If, for instance, white men began competing with black men whom they considered overly pushy in the pursuit of white women, then the word *jealousy* would raise its ugly head. If, in a university, students competing for the few doctorates available complained that some individuals were being overly pushy, they might say those people were aggressive go-getters. Could it not be that the university students were jealous of one another and that the black and white men competing for white women were simply competitive aggressive go-getters?

The fact is that many words are used exclusively for political or purely selfish reasons, not for trying to communicate objectively. Some of these words may be correct in some instances and wrong in others. It may be a matter of opinion to where these words belong.

Euphemistic Words and Phrases	Put Down Words and Phrases
Fighting	Squabbling
Extreme fighting	Going berserk
Fighting one's own battles	Taking the law into one's own hands
Asserting oneself	Acting like a spoilt child
Taking it on the chin	Taking things lying down
Pointing out an injustice	Winging
Criticising greed, lust, or conceit	Being jealous
Being firm and resolute	Being stubborn
Reasoning things out	Being weak and ineffectual
Thinking things through	Being indecisive
Being decisive	Jumping the gun
Acting quickly and decisively	Getting in a panic
Keeping cool and calm	Being half asleep
"He's a man"	"He's a chauvinistic pig"
"He's a man"	"He's a bully, a barbarian"
"He's a new man"	"He's a wimp"
"He's a gentleman"	"He's a softy"; "One has to be a man first before one can be a gentleman"
"He's civilised"	"He's afraid of barbarity or of being a man"
Retreating	Running away
Sticking up for oneself	Cry for help to the people around oneself in an underhanded, Machiavellian way
Sacrificing	Giving up the ghost
Standing on one's dignity	Acting like a pompous prat

Sometimes, the up word or phrase is relevant to a situation; at other times, the down word or phrase is. And sometimes, it's simply a matter of point of view. I think it's about time that we defined these words and phrases logically, creating precise definitions. The worst misuse of words is when the speaker includes a subliminal message in order to assign a person or thing a false value. During the sixties, some Labour Party Members of Parliament (MPs) were fond of using the word *rational* (pronouncing it "RASHnl"), emphasising the "rash" part of the word. The word *rash* in English means "headstrong," i.e. "to charge in where angels fear to tread," which is opposite to the meaning of the French word *rational*, which means "the ability to reason things out logically and calmly." It was as though the MPs wished to put across two apparently conflicting points. I am a reasoning person, but, at the same time, I am not dull and boring. I am a doer who gets things done. I have the guts to charge in where angels fear to tread.

Also, the word *raze* is a French and a Latin word used to denote destruction ("The building was razed to the ground"). Is it just a coincidence that there is also a like-sounding English word, *raise*, which means "to rise up," as though one is being constructive? Some people use this word to fool the listener or, perhaps, themselves.

The term *human being* is used by people who wish to ascribe high value to a member of the *Homo sapiens* species, no matter what his or her behaviour. Is it coincidental that the term *human being* sounds similar to the term *humane being*, which means "a person of great compassion and dignity; a person of high value"? It's as though a person who uses the term *human being* is trying to tell someone else that the person in question is a member of the species, a fact that the ego is forced to accept as true. But at the same time, the self judges the word not by logically evaluating it but by evaluating it emotionally. The self cannot analyse in the same way as the ego. It is therefore forced to take on the emotional baggage of the word, the meaning "a human being; a person of compassion and high value" since the self can only evaluate words emotionally. Therefore, when someone describes another as "a human being," the audience is subliminally "brainwashed" into believing that the person in question is of a high humane character – even if that person is a cold-blooded murderer. You can see now how people twist and distort language. A similar method is even used in some courts of law. I am sure that if it were understood that twisting and distorting language is a common technique to "brainwash" people, doing so would not be permissible in a court of law.

I mentioned survival early on in this book, saying that it is a relevant issue today, whether one is competing to overcome the very hard physical problems that coming with living and surviving in the environment or competing with others of one's own kind for the limited but easily available resources. If one individual in the hunt is better at pulling down the prey than the others in his pack (this ability may be due to swiftness of foot or skilfulness in making and using a weapon), then it is to the long-term advantage of the pack to see to it that this individual has first turn at feeding on the kill so that he is kept fit for future hunts as well as future breeding (as his genes will contribute to the pack's genetics-based physical and mental health. However, if the pack members gang up and do not let him have the advantage of the kill, they may profit in the short term but not in the long term. If the fit and healthy have more young, then there will be more fit and healthy young for one's own few young to interbreed with in future. In this way, the pack will survive better.

If one lives on resources that are easily available, then the odds are that one competes directly with one's own kind. In this case, the ones who survive are the biggest, strongest, and cleverest (provided that each individual competes for resources on a one-to-one basis). However if the process of competition were based more on who was the best Machiavellian in ganging up with others, then the most Machiavellian would win out, but not necessarily the strongest or most academically intelligent.

Successful people do not necessarily make successful societies. It is a known fact that cancer cells are very successful in the human body. They absorb plenty of the body's resources by seducing the body into laying down plenty of blood vessels to feed the tumour. Cancer is, therefore, "job-creating." Cancers are not normally killed by the immune system, because the immune system only looks for foreign or weak and unhealthy cells, which the cancer cells are not. In the end, the cancer is so successful that it kills the body. So it is with Machiavellians.

If it is true that people who reject Machiavellians survive better in the long term than people who do not, then one might ask, "Why are we not conditioned by natural selection to reject the Machiavellian?" I would suggest that we are. When we see someone like Yasser Arafat, a classical Arab, are we not repulsed by his "friendly" manner? (Think of how he greeted Saddam Hussein on TV.) Arafat is a Machiavellian in Machiavellian clothing (I am glad I was not having my breakfast at the time of that broadcast). Unfortunately, not all Machiavellians will repulse everyone in this way, as the manner of some may be more subtle.

There is, of course, the usual arms race in all aspects of evolution. When one response evolves, an evolving change in the organism appears to counter that response. Not all Machiavellians are as obvious as Arafat. Therefore, we must be on our guard, especially since the politically correct shame people into ignoring their natural instincts. This is to say that much of the revulsion we feel when witnessing certain types of behaviour may simply be programmed into us by evolution so that we keep clear of the Machiavellian, but we are brainwashed by the politically correct into believing ourselves prejudiced and bigoted.

The British economy expanded during the 1980s. The reason usually provided for this was the emergence of the yuppie. If yuppies were ingenious engineers who invented new products, found better ways of making things (cf. Sir James Dyson), or made marketing better by offering the right design at the right price (cf. Alan Sugar), then I would be 100% for the yuppie generation. But this, unfortunately, is not the case.

To analyse production, I offer the following questions and opinions:

If a man wins the pools, then would you say he is a productive part of society?

If a person wins a bet on a horse race by using their intelligence to study the winning horse's form, then are they being productive?

All that these people are doing is redistributing money from the many to the few. They are producing nothing.

Of course, a person could employ people to help them study form and then say that they are creating jobs. Yes, but the jobs they create are no more productive than they are.

If a man buys shares of stock and sells them a day later at a profit, then has he produced anything or contributed to the economy? (If a man puts money into an industry to buy machinery for future production, then that is a different case, since such a thing would create wealth.)

If a person buys goods one day and sells them the next, have they produced anything? (I would agree that a distributor may be useful to distribute the goods around the country, but this would be thanks more to the lorry drivers than to the shopkeepers, although I would agree that without shopkeepers, things would be very awkward for society.)

I am not suggesting that shops and other services are a waste of time, but only that they reveal the symptoms of a thriving economy that the shopkeepers and service providers did not create. I will give an example. When Sock Shop, which sells exclusively socks, sprung up, it did not improve the economy, for the simple reason that if one shop sells more socks, then other shops sell less. Of course, one could say that people bought more socks after Sock Shop opened, which encouraged trade. But if people didn't buy socks, they would have bought something else. Therefore, for every person employed by Sock Shop, another person is made redundant elsewhere. There is only so much money in the economy.

If, however, you make goods cheaper by using automation, then you do not have to reduce workers' wages. You can still sell the goods for less and not take money for buying other things out of the economy. Also, you may encourage people to buy domestically made goods and fewer foreign goods, helping the balance of payments and preventing a loss of jobs to industries abroad. Inventing new products and automating the manufacturing process, not sales gimmicks, are what help strengthen the economy.

Some people have the idea that automation creates unemployment. This is only true in the short term. Let's look at a hypothetical example of what happens in the long term. Say that a TV manufacturer makes one TV per employee per day. Ten workers are employed by the manufacturer. Suppose the manufacturer develops or buys a machine which produces ten TV sets a day and needs only one person to operate it. In the short term, this would mean sacking nine employees. But then suppose that the manufacturer bought ten machines and employed a person to operate each one. The manufacturer would then use all ten of the original employees. The manufacturer could then produce ten times as many TV sets with the same number of employees. So, it is faster and cheaper, opening the possibility for the employer to pay higher wages. Because the TV sets were cheaper, the manufacturer would sell more at home and abroad.

To help clarify automation, I'll say that if a society produces more goods per person per hour, then that society can consume more goods per person per hour. Some people judge their own wealth by looking at how much they can consume to raise their standard of living. I do not believe that one can judge the economy in the short term by looking at how many businesses there are or whether there is full employment. I believe that the gross national product is the most important sign of the economy's health. One should not go blindly looking at short-term, bottom-line figures. The fact that the British economy experienced a balance-of-trade crisis the moment the economy expanded in the late eighties proves that it was only high unemployment (and therefore a shortage of demand for goods) and North Sea oil that kept the balance of trade in check. As soon as heat is applied to the economy, everything falls apart. A genuinely strong economy would not fall apart in such a circumstance. In the late nineties, the British economy improved, perhaps because the other European economies fell behind.

The reason why Britain has a great divide between the rich and poor (The British poor are nowhere near as poor as the poor of the third world "as yet" please note) is because most of the investment is in property which is "static" wealth and not in production "active" wealth. The property wealth only helps the wealthy, but manufacturing wealth helps everyone.

The question that should be asked is why the wealthy, "the powers that be", do not give a dam about what happens to the poor or the rest of the population, maybe they do not like, or identify with the population as a whole. Maybe they do not like the population because, dare I say it, because the population have become fat and ugly (not making the best of themselves) like a population of Quasimodos. (This I suggest is the subjective perspective of the powers that be of course.)

Introverts and Extroverts

I have heard that there are two types of people, extroverts and introverts, which idea I have always seen as rather simplistic. I would say that whether a person is extroverted is determined by seeing if a problem that a person is to solve can be solved according to his or her nature. If a person is an ego type, has an individual problem to solve, and has plenty of natural resources to solve it, then that person would be very positive, and therefore extroverted, towards the solution – just as a self type of person would be positive if the problem were one of seduction. Both types would, however, be somewhat negative if they had to deal with the opposing type's problem. For most people living in a civilisation, using seduction is very important in getting a job. Since we tend to judge people by their social behaviour, it is quite obvious that the self type, who is at ease in a social situation, would be judged the extrovert, whereas the ego type would be judged the introvert.

Chapter 17

Fighting, and What Is the True Nature of Humankind?

Fighting

The term *fighting* is often used in Great Britain, and yet people hardly see fights. I suggest that there are two ways of fighting.

The Ego's Way of Fighting

The ego's way of fighting is like the way of a knight in shining armour who battles with a dragon (in archetypal form, that is). The ego's way involves danger and true guts and courage. If it is something in particular that the person is fighting over, e.g. treasure, a maiden, land, etc., then that something has no choice in the matter. Whichever opponent is the best fighter in the objective sense will win and claim the prize.

The Self's Way of Fighting

The |self's way of fighting includes the instance of two women fighting over the same man, even if the two women never actually meet, e.g. a wife and her husband's mistress. It also includes the instance of prostitutes' fighting over a john, trying to outbid one another and determine which can give the most or turn the best tricks for the same money – or the same tricks for less money. With a self-based fight, there is always a third party involved who sits back and laughs his or her head off. This person may have already made a choice between the two prostitutes but is hoping to get a better bargain. The more the self fights, the better it is for the third party, since the third party can demand more and give less in every way.

The third party can invent opposition where none exists so that the person in the position of the self decides to fight even harder to make a better deal. Some husbands invent the idea that they might have a mistress so as to manipulate their wives into being more amorous and submissive. Also, it isn't unusual for a salesperson to invent customers who are after a product, which, of course, is in "high demand" and "short supply," in order to make the customer less sales-resistant. Is it not possible for an employer to invent people after one's job, someone who would do the job faster and for less money, in order to make one frightened of losing it? Therefore, one will work harder for less money. The ultimate insult is to be called a coward or weakling because one is not "fighting" (competing) hard enough.

Note that the fight of the ego is noble, proud, dignified, and rewarding. It is a measure of how high one can ascend. However, the fight of the self is degrading, submissive, and undignified (to a man, anyway; I cannot know how a woman feels). It is a measure of how vulnerable one is and how low one is expected to sink. When fighting for a job, pay rise, or promotion (or anything else that requires permission), ask yourself which type of fight it resembles, the ego's or the self's. Be it far from me to judge. Please make up your own mind (the author says sarcastically).

When I hear that there are many unemployed young black men in the streets, I have every sympathy for their position. They know they are men within a society that asks them, "Why don't you be a man and go out and 'fight' for a job?" I do understand the anger and resentment that must burn inside them. They know that, being black, they have to "fight" (crawl and grovel) for a job far more than an arse-up white man would have to.

Young black men in the streets are also aware that one does not "fight" for a job, although, not having an academic background, they may not know how to argue this out. Civilisation confuses the meaning of "fighting" with "seduction", which must hurt them more than they hurt a white man. Their manhood is very important to them. It's the only thing many of them feel they have got.

To make a point: A person going through a period of labour knows that the product of his labour belongs to the master. The person is to recognise the place under the master. If the person does not like being under the master, then the person is supposed to fake it; otherwise, the master will be offended. People in this situation are not to make decisions unless the decisions are so unimportant that the master does not care one way or another. The person is given just enough money to keep him "happy," meaning "healthy," so that the person will labour some more.

Is the above paragraph to do with the relationship between a traditional (Victorian) husband and wife or between an employer and employee? If you cannot tell, then why not?

The True Meaning of Fighting

Tom, Dick, and Harry

> This is a hypothetical story. There were once three solders, Tom, Dick, and Harry who were caught up in a war and were lost in battle. They had heard that the enemy would shoot one of every two prisoners. So, when the three solders met the enemy, Dick and Harry threw themselves down on the ground and competed with one another over who could crawl and grovel the most in order to save their own skins.

They grovelled, crawled, whimpered, and begged, saying that they were human beings and that the enemy had no right to shoot them.

Well! Tom stood back, amazed by the behaviour of the two on the ground, and said to the enemy, "Look, don't expect me to do that. I would rather be shot. If you want to shoot me, well, that's up to you, isn't it? You can shoot me if you want to." None of the three soldiers were shot. It appeared to be a misunderstanding arising from propaganda that one of every two prisoners would be shot. All three soldiers were repatriated in their own country. However, there was an inquiry into the behaviour of the soldiers in the face of the enemy. The three soldiers were brought forward and asked to describe their own and each other's behaviour.

Dick stated the following: "Well, sir, we were told that the enemy shoots some of its prisoners, so when we met the enemy, we thought the worst. Harry and I, being the competitive type, realised that we would have to fight for our lives. Well, I mean, we are fighters who are not afraid to fight. Understanding that this is a dog-eat-dog world, a man has to do what a man has to do. We were not afraid to express ourselves. On the other hand, Tom just stood there in the face of the enemy and said, 'You can shoot me if you want to.' Well, the man is obviously a coward. I mean, he is even afraid to show weakness. Well, my mate and I are not cowards, as we are not afraid to show weakness. We stuck up for ourselves and told the enemy that we were human beings. We are not afraid to knuckle down and do what a man has to do to survive. We are doers. We did something. We didn't just stand there like wimps and say nothing but, 'You can shoot me if you want to.' We are survivors. We see life as it is and face up to it. We don't take things lying down. Tom is the sort that lets the world walk all over him, the sort that takes things lying down. I mean, fancy just standing there and saying, 'You can shoot me if you want to.' He's a quitter, the kind that gives up the ghost, unlike us, who are fighters."

You can see that Dick's statement is a lie of perception. But, given the way that words are used in Western Society (which is in a distorted fashion), there is an element of verbal truth in Dick's words. But if you saw the original scene as described, then you would understand that what Dick said is a total and utter lie. The fact is that this type of situation occurs time and again in western civilisation, although not quite so blatantly. We all know that we do not "fight" for a job, no matter how we interpret fighting. It all comes back to the self's interpretation of fighting. The self does not understand the ego's concept of fighting. It only perceives conflict, struggle, and competitiveness; therefore, it interprets any competitiveness or extreme effort as fighting. Since it perceives from the ego that fighting is supposed to be brave, it therefore evaluates its own actions as brave, forgetting, of course, that having bravery or courage is to face death or danger, not to submit in the hopes of avoiding death or danger.

I have heard it said that if a man or woman who "acts" bravely then he or she is brave. I think sometimes people confuse "acting" as a person would act on a stage, with that of a person who "behaves" bravely. That is to say puts his or her self in a position of danger for a good cause.

The True Nature of Humankind

As far as the social sciences are concerned, and as I have suggested before, one has to use different criteria to come to a conclusion about anything. When dealing with the true sciences, though, the behaviour of water, for example, is not changed when one reveals the facts about water. We consistently know where we stand in relation to the structure and behaviour of water (or any other physical property). However, when dealing with the social sciences, as soon as facts are revealed about humankind, it is very possible for people to change their behaviour and prove the social scientist wrong, particularly when people are offended by anything that is uncomplimentary to their nation, race, religion, or way of life. I as a British person will always see the other's point of view and answer criticisms accordingly, sometimes I will totally agree with them. There has been times that British immigrants have been criticised by the Australians. I have found the some of the criticisms very constructive and somewhat amusing which has changed certain behaviours in the Brits over there

When sociologists talk about human behaviour, they are trying to assert facts from a state of ignorance. As a very young man, I felt rather self-conscious and a bit of a wimp because I had rather a thin physique compared to my peers. I have worked hard in building a bigger physique. However, this does not mean that I am a natural athlete or enjoy sport. Far from it. As a boy, I liked climbing trees and playing with bows and arrows, etc. Sport, on the whole, bored me stiff. Now, my bigger-than-average physique is a divisive characteristic I wear, not a natural one, although it does reflect that I like to be, and feel like, a man. The fact that I had to work hard, harder than a natural athlete, to acquire the same effect was simply because one normally has to work harder at something that is not natural. Had I been surrounded by boys the same as myself, I would not have known any different and, therefore, would not have done anything about my physique. The same could be said of some people in academic or intellectual pursuits.

As a small child, I, at times, wanted to run about playing soldiers or cowboys or making a noise to a greater or lesser degree. My mother and grandparents would often ask me, "Why don't you sit down quietly and draw or paint a picture like your sister? There's a good boy." Reluctantly and after much persuasion, I would "be a good boy" and start painting, with a certain resigned feeling of boredom. This friendly persuasion did not turn me into an artist or develop any desire to be an artist. The fact is that I loved, amongst many things, to make paper aeroplanes, which I loved to glide. By using trial and error, I'd find the plane that would glide the furthest, much to the total disinterest of the people around me. When I showed my superiors the product of my "hard labour," they were too busy to express any praise or compliment. However, this did not destroy my interest in the sciences. The point I wish to make is that although I first had to know about aeroplanes, what I did next, out of choice, was a reflection of my true nature, not simply cultural encouragement.

One must be careful to differentiate between what is a natural and what is divisive. It may be the case that some people throughout the world may be happy to wallow in religion and emotion. If a Western person were to say to them, "You are inferior, as you do not know anything about the sciences" (whether or not a person is inferior and knows little about science is not the point, of course), then they might study science and work hard (harder than a person who had a natural flair for science) to earn a degree, only to prove to other people that they are not inferior. In this case, the interest in science is not a natural but a divisive one. The important point here is that an individual like this only alters his or her behaviour because the Western person is looking on. If the Western person did not exist, then where would be the natural progress? I believe we all know people who worked hard to pass their exams but then never bothered to educate themselves after that point, holding the belief that they have proved their intelligence (their worth). Thereafter, they sat back without any interest in intellectual pursuits whatsoever.

These people are not natural intellectuals. They use intellectualism as something to wear on their sleeves, to indicate an identity or to "pose." It only proves that they are prideful and wish to appear intelligent. These people are very good at remembering facts and figures, and they tend to study by conforming slavishly to the established learning principles, regardless of whether the information is of interest to them or would be useful in the world outside academia. Some breakthroughs in human achievement are made by people on the fringes of the scientific community. These people may not be specialists in one specific field.

Throughout European history, humankind has held the desire to progress. In order to progress, a culture needs time and resources for experimentation. In every society, there will be wealthy people who have both time and resources at their disposal. Now, it is said that in the Third World, people live in a feudal state and are either rich or poor. If one is very rich, then one does not need to progress, because one is happy with the way things are. Therefore, there is no incentive to progress. (Being rich does not necessarily mean that one has gold, silver, or currency, but one does have manpower at one's disposal. One cannot say the rich people in the Third World are short of manpower.) If one is poor, then one doesn't have the time and resources to progress. In Europe in the Middle Ages, people lived under a feudal state and achieved some progress. The Industrial Revolution was started not by poor engineers but by very wealthy men who had a natural interest in science and engineering. Why didn't those men sit back and enjoy their wealth like the wealthy people in the Third World? I would suggest that there is something more than circumstances that shapes a culture. It is something about the *nature* of a people. That nature is evident in the rich people of a race or nationality, as they are able to express their nature more freely.

To some it all up, I say that we must evaluate what state the human race is heading for. Is it a universal Third World state, as Western people interbreed with and are out bred by people of the Third World, and civilisation ends up as a giant, universal Brazil, with a few rich people and a mass of extremely poor people?

It appears that there has been far less real progress in civil aviation over the last twenty years than in the twenty years prior. From World War II to the late sixties, commercial aircraft, in both size and speed, jumped by leaps and bounds with the introduction of the jumbo jet, the Concorde, and the fighter Sea Harrier. But since the sixties, there has been little progress, except that bigger aeroplanes have been manufactured. Of course, there has been progress in electronics, but this is because there is so much profit in them. Made by mass production for a mass market, electronics require only a relatively few engineers, plus a lot of people to do the screwdriver jobs in the process of manufacturing. With aviation, up until now, that is, the people building the planes need to be highly skilled. But when new methods are found to turn out planes like sausages, cheap Third World labour will take over. Cheap Third World labour may mean British people in the future.

In the automobile industry, there has been a revolution in car design, emphasising safety and convenient gadgets rather than any revolutionary change in engine design. Today's cars retain an internal piston engine, which would easily recognised by a First World War engineer. I wonder if this is a sign of things to come. People who predict the future overestimate humanity's future progress. I do not think it is simply a case of over-optimism. I think it is that the humankind's progress has slowed down as a result of Machiavellianism.

Also, the Machiavellians have gradually encroached on the Western world. I think that the human race is heading for disaster. Humanity will recover when the ego asserts itself and people will return to progressive production rather than playing second fiddle to seductive Machiavellian bigots.

Chapter 18

Control, Power, and Poverty – and a Summary

Many people may feel powerless in this world in which we live – powerless, even, to control their own lives. There are many victims in this world, such as blacks, females, disabled people, deaf people, old people, homosexuals, people of any kind of ethnicity, et al. (I do not believe that people who are addicted to alcohol, other drugs, or gambling are victims, even if they claim to be.) This does seem to be the age of the victim. I myself am a victim because I am a middle-aged, heterosexual, Anglo-Saxon, healthy, and able white male. I am a victim because I do not have a victim's status. Well, I have been out of work, but now I am retired (which hasn't made me feel like either a victim or a loser, I might add). In fact, having a job and all the pressures of trying to keep it may make a person more of a victim than when he or she is out of work. (When I was out of work, it gave me the time to think and the opportunity to write this book.) I recently read that young people who are out of work live with greater hope for the future and feel more at peace with themselves than those who hold low, menial jobs which they feel are beneath their talents. Because of my own personal experience, I was not surprised to read this.

There are two types of problems.

With the first kind of problem, one must ask the questions that follow: "Do I have to ask someone's permission in order to solve it?" "Would I need to degrade myself in order to solve it?" "Do I need more money than I have afford in order to solve it?" If the answer to all three questions is no, then, within reason, that problem is more likely to be under one's own control to solve.

The second type of problem can only be solved by asking, and getting permission from, someone else. In this case, no one has the right to say that finding a solution is entirely up to you.

Let's analyse, Does one need permission to borrow a book from the library, to buy a paperback book, (assuming that one can afford it), to read a book, to learn, to think, to live within one's means (which means not taking on any commitments that one does not really need in order to enjoy a present-day lifestyle), to be without vices (gambling, using drugs, drinking), or to remain single or childless? Obviously not. One cannot blame anyone but oneself if one fails to do any of the above.

Does one need permission to get a job, a pay rise, or a promotion? Obviously, yes. One cannot blame oneself if one's problem is lacking the qualities to get a job, a pay rise, or a promotion. Even if one has many qualifications and an IQ of 1,001, job interviewers have the right and ability to deny one a job. Also, if one can get a job just by being the right man or woman for it, then why are there laws against race and gender discrimination?

Choices

When we are offered choices, do those choices concern circumstances and the workings of the physical world and are therefore within our control, or are those choices devised by people who use and manipulate others for their own selfish ends? Are these people asking others to *jump through hoops?* A choice is not a choice when one is damned if he or she does and damned if he or she doesn't.

What's Fair?

I learnt at an early age is that life is not always fair and can never be guaranteed to be fair. One problem with this world is that too many people expect life to be fair. If you look around the world, then you will see people who are worse off than you. Many of the people who shout the loudest about the world's being unfair are those who have the type of problems for which they need not ask someone else's permission to solve. The only type of fairness that we should demand is fair play when it comes to the continuous progress of civilisation, considering what is fair to society as a whole. Where individual rights are concerned, I am all for the right of the individual to be left alone and not to be victimised, regardless of race, culture, or beliefs, and the right to express an honest opinion, no matter how much in a minority any one person may be. I would also add that if we are to be criticised, then we have the right to straight-talking without any Machiavellian, word-distorting double-talk.

With regard to economics, countries often talk about a "level playing field," disapproving of countries that use subsidies to keep their markets open. People in the United States complain about European farming subsidies but don't like it when it is pointed out that United States industries get information from the NASA free of charge, whereas other countries have to pay for it. The West could complain that "slave labour" in the far East undercuts Western manufacturing. Also, the United States gets its oil cheap (when compared to the price Europeans pay for oil) from American oil wells, which could be considered a form of subsidy. The Americans could say that their cheap oil reflects a natural market price, since it is cheap to produce in that area and the distribution costs are low. In other words, it is not divisive. But then it could also be said that cheap labour is also to do with the market price in a certain area. Exactly where does one draw the line?

In warm Third World countries, people do not need to spend money to keep warm in the winter, as is the case in Northern Europe. People in the Third World, therefore, can survive on lower wages than people in the West. Also, the land is more expensive in Europe than it is elsewhere. European industries have to pay for their land. I could go on and on about Europe's industry being disadvantaged and not on a level playing field. I cannot say for sure which industrialists are is right or those that are wrong, for the answer is far from simple.

Poverty

In modern societies, how does one define poverty? It is obvious that poverty is relative to the society in which one lives. But there is more to it than that, since many people who could be described as financially poor do not feel poor and would not even describe themselves as poor. I believe that one cause of poverty has to do with power, not power over other people, but power over one's own life, the power to make decisions for oneself. This also entails power over one's own territory, which includes personal security and personal privacy. If you look at various cultures that have different degrees of wealth, then you will inevitably see who is rich and who is poor, even in a primitive tribal encampment.

The rich have more personal space and protected privacy. If one wished to make an attack on the rich, then one would have to go more deeply into private territory, which would make one, instinctively speaking, the "Perpetrator" of a legal and moral wrong. Anyone who has read Robert Ardrey's books, specifically *The Territorial Imperative,* will recognise the following concept. When an animal intrudes into another's territory, it grows more nervous and apprehensive the deeper it intrudes. The closer the intruder comes to the centre of the other animal's territory, the more aggressive the defender becomes. The intruder, at this point, becomes more willing to retreat, until the intruder is expelled. The law seems to take this into account, presuming that there is a consensus on the point that the deeper the intrusion, the greater the defender's right to retaliate – and the less right the intruder has to attack. If a person has very little personal territory, then he or she can be far more easily attacked, without too much legal and moral intrusion into his or her territory. Of course, if the intruder has to destroy structures in order to intrude, this serves to create at least the illusion of territorial distance. Therefore, by a combination of circumstance and instinct, I say that the law gives far greater protection to the rich than it does to the poor.

There is, at the moment, an attack on the welfare state in Great Britain. The idea is that people should stand on their own two feet. When one lives in an urban civilisation, it is hardly possible for one to hunt animals in the forest or cut down trees to use as firewood or for building a house of one's own. Many urban-dwelling people are, given the nature of civilisation, interdependent upon one another. People who live in urban or suburban areas are often dependent on other people's building their houses, supplying their fuel, growing their food, and so forth. If one has money, then one can secure the services of the people upon whom one is dependent (some people cannot afford such services). It is all very well to say that one should pay one's way, but the very fact of paying one's way demands that one rely upon others to give one the means (e.g. a job to earn money) to pay for the services. In such a case, one is still dependent on society, whichever way one looks at it. What I would say is this: An individual cannot help what talents he or she has or has not been born with, nor can one help the wealth or poverty that one is born into, but one can help whether or not he or she marries or has children. If all people who found they could not make their way in life (because of a lack of talent, for example, whether or not it was a person's own fault) were to remain single and childless, then the cost of maintaining a welfare state would be minimal compared to what it is at present. By choosing not to marry and have children, underprivileged people would make it easier for the government to help them, which would result in more other underprivileged people's rising out of poverty. If the government becomes alarmed when too few babies are born, then it knows what it can do. If there are a great deal of talented unemployed people whose talents are not being exploited, then I think the British or any government should have a word with industrial and commercial executives, not whinge to the unemployed whose power over industry and commerce is non-existent. As I have already said, one cannot take responsibility for something one does not control. To say that one could seduce, persuade, or lick arse is to forget that the ego, which engages in efficient production (which keeps Western civilisation going), does not have the characteristic nature to lick arse to make such things possible. I would therefore suggest that these behaviours are better for self-dominant Machiavellians not for ego types.

The Power and the Glory

It is my experience that men who are in a position of power see themselves as natural leaders, or alpha men, as though this were an expression of their own genes, as though their power over others and the ability to make other men cringe in fear was only incidental to their positions of power. I, for one, have always seen power as something that is awarded or given for a person's long service, inside knowledge, or intellectual ability to do the job. But because of Machiavellian manipulation, this is not as a reflection of a natural alpha.

I am somewhat amused by women's rise to power in the marketplace. Now, some women who have risen to the top of some companies impose their newly gained power on the men around them with, I have heard, a humiliating vengeance. I wonder if the men who are now reduced to tears in the boardroom after the feminine onslaught are the same men who would believe that a position of power reflects a true genetic alpha tendency to be a natural leader. Do they know that conceding power is not now the natural thing that they thought it was in the past? Have they discovered that one who holds power should also hold a responsibility for treating people well? If a white person reduced an ethnic person to tears, then there would be hell to pay, wouldn't there?

Law and Order

People of the legal establishment believe in the word of the law and people who stray from the law or take the law into their own hands should be punished. The British judiciary, seam more concerned with the book of law than with the public's welfare. I am not suggesting that laws are not necessary. On the contrary, people need to know where they stand, how to deal in business, what taxes to pay, how to follow the Highway Code, etc. It is just that sometimes the judiciary forget why we have laws. Laws are made so that civilisation may run smoothly. Laws are not to protect criminals. They are to protect civilised people. Lawyers exist for the purpose of protecting those civilised people who may be wrongly accused of breaking a law and also civilised people who are prosecuted under bad laws created by an administration.

Laws and lawyers should not exist to protect the guilty, who do nothing but show contempt for the general civilised public. It is as though to a lawyer, the law is a religion whose tenets should be unquestioningly obeyed without reason or thought given to right and wrong. The way in which the judiciary thinks reminds me of the film *The Magnificent Seven*. In the film, Mexican peasants are victims who need someone to protect them from bullies. The peasants are only too happy to pay what little they have for protection. They see the Magnificent Seven as a "shining light," as people who would protect them. The Magnificent Seven, however, are only paid in rice, please note. The important thing to understand is that without the bandits, the Mexican peasants would not have been victims. Without victims, there would be no need for the Magnificent Seven. Therefore, bandits and victims are the Magnificent Seven's bread and butter, or bowl of rice.

By using deduction, I state the following. People who practise law are necessary because criminals and their victims exist. Without victims, most people who practise law would simply be a waste of space. If someone came up with a solution to the criminal problem, then this would mean that most people of law would find themselves out of a job, out of pocket, and out of their shiny, "up in the clouds" lifestyle. Is it any wonder that the criminal problem is out of hand? I am not saying that people who practise law consciously want society to fall to crime, but only that there is no direct incentive for them to avoid feeling this way.

When the issue was nuclear deterrence, the establishment was only too aware of the fact that so long as the enemy knew that the opponent could and would do drop a nuclear bomb, it would not be necessary to do it. Even though dropping a nuclear bomb meant killing innocent people, i.e. women, children, and non-military men, by way of nuclear radiation, this did not prick the establishment's conscience. However, the idea of putting the majority of dangerous criminals to a quick death would throw the judicial establishment into spasms of self-righteousness, indignation, and condemnation. The atom bomb made its point to the Japanese leadership who would otherwise not have surrendered so quickly, and would have cost a lot more lives in the process.

The reluctance of the courts to reveal past criminal activities of the accused while on trial bothers me. Many a jury has given the benefit of the doubt to an accused person, only to learn afterwards that the accused had an arrest record as long as his arm. Now, there are instances where evidence is put forward in such a way that it is remotely possible for a jury to see the accused as a victim of a series of unfortunate circumstances. No jury wants to convict an innocent person. However, after considering the past record of the accused and the circumstances involved, it sometimes appears totally obvious that to acquit the accused would be stretching the coincidences a bit too far. If a lawyer says that a defendant should be judged only by the evidence for the crime of which he or she was charged, I would say that the defendant's criminal record is a relevant piece of circumstantial evidence concerning that individual's character. It is as important as finding the defendant's motive for committing the crime or as knowing his or her state of mind at the time. Denying admission of a defendant's criminal record leads to a situation that is rather like trying to make out a picture on a computer screen by seeing only a few pixels at a time. Only by seeing as much of the whole picture as is available is it possible to get near to the perception of the whole truth.

The general reason that is given for the existence of the judiciary is that the judiciary protects people from criminals. I would say that just the opposite is true. If there were no law and order and if criminals thought they could do what they liked, then the general public would retaliate in no uncertain way and kill the criminals. I would suggest that the judiciary protects the criminal – or those accused of crimes – from the public. The judiciary, however, are very good at crime solving when the public do not know who committed the crime. I would suggest that it might be better if the people who practised law determined what the accused did according to the complex evidence at hand and if the jury then decided the punishment. When the issue at hand is moral in nature, people of the general public are just as qualified, if not better qualified, than the people of law to present the case. If every criminal were given a death sentence and then had to rely on the victims or the victims' next of kin to make a plea for mercy on the criminal's behalf, then criminals might have a different attitude towards their victims other than showing the contempt they normally express.

I remember seeing a TV programme about the criminal philosophy of the utilitarian. A philosopher came up with the idea of executing people who parked their cars badly, saying that badly parked cars hold up ambulances on their way to hospital, which leads to many lost lives. A society would only have to kill a few car-parking offenders each year, because most people would be deterred from illegal parking. Compare the loss of a few car-parking offenders to all the lives that would be saved if ambulances got people to hospital on time. Very good point, I must say. However, when all is said and done, I think it is a matter of people's choosing what sort of society in which they wish to live. I would personally weigh up the choice between having an accident and not making it to hospital on time and being put to death after wrongly parking my car. I would, of course, choose the status quo.

The main point here is that this choice is not a matter of weighing so many lives, like so many heads of sheep, but a matter of what a population is happy with. If a population lives in fear of the law and of a police state, then it is obvious that the law is overreacting to crime. If a population lives in fear of criminals, then the law is under-reacting. When all is said and done, it should be up to the population as a whole, not philosophers or scientists (social or otherwise) to come to a conclusion about law and order – whether to enact the death penalty and which sort of criminals deserve capital punishment.

Times Past

It is said that, in times of social unrest, people tend to look back with nostalgia on a more peaceful and less violent time. Pundits then say that if one looks in the newspapers of times past, one finds that those times were far from peaceful. I think the main problem here is that the violence that occurred in Great Britain in the past existed exclusively in the "rough" working-class areas and did not affect the rest of society. Also, in times past, if a man committed a crime, he knew he was doing wrong and did not have contempt for the law. Youngsters knew that if any adult shouted at them and told them off, they were not to answer back or give any cheek. If they knew they were doing something wrong, then they simply ran away. Children, in times past, respected every adult and were under threat of punishment by their own parents. My own grandmother, who lived in the East End of London, could leave her door unlocked and go shopping without any fear of being burgled. Nowadays, many children have lost all respect for adults, the police, and even the law itself. In times past, it is said, British society tolerated more crime than it does today, but this takes into account only crimes of violence between two consenting adults during a violent argument, which occurred mostly between rough working-class people. This does not mean that society tolerated a ruffian who physically attacked a respectable person or robbed someone. On the contrary, if anyone living in the fifties in Great Britain were found guilty of theft, assault, or murder of an innocent person, then he or she would be punished far more severely than he or she would be today.

On Euthanasia

I have often wondered why doctors are against voluntary euthanasia when patients are slowly dying and needlessly suffering, having no chance of being cured. Why don't people have the right to take their own lives when they find that life is not worth the living? Euthanasia and suicide are illegal in Great Britain because, just as lawyers need criminals, doctors need victims of bad physical or mental health. I consider that all professionals guard their own professions as though they were the centre of the universe and life could not go on without them. I restate my opinion: The more victims professionals have, the stronger they become.

*   *   *

Throughout this book, I have given my uninhibited point of view on life, so that the reader may see life from a different perspective, one that is not imposed by the establishment. A reader might say that I express old ideas and provide explanations that are different from those of the past. I doubt that I will be able to change the world by offering my point of view, but if enough people in positions of power read this book and agree with some of the ideas herein, then, who knows, what will happen. I will always remember the Hans Christian Andersen story "The Emperor's New Clothes," as it sums up exactly what I think and how I feel about the world and the society in which we Britons and the rest of the world live.

A Greek philosopher once said, "People do not *think*; they simply follow *trends*" (emphasis mine).

How times have changed!

I remember that as a child, in all my innocence, I would say something quite openly and then be rebuked for being wrong. I remember saying something like, "I think that's ugly," and then being told that beauty is in the eye (mind) of the beholder. Well, OK, I can understand what that means, if it's a matter of evolutionary response in a species. But if it is true that nothing is beautiful or ugly, then why aren't film actresses chosen randomly to play lead roles? Why do people object to a motorway across green fields or an industrial site at the edge of their gardens? Isn't the motorway or the industrial site merely reflective of the ugliness in the mind of the objector? When a person is disfigured in an accident, then why does he or she or the state spend so much money on plastic surgery, to satisfy all the "ugly minds" out there?

Question: Does ugliness have to exist in order for one to know beauty, given that ugliness is the absence of beauty? I believe that things can be neutral, that is to say, neither ugly nor beautiful. I do believe that people can get rid of ugliness if only by replacing it with neutrality and beauty. However, if ugliness needs to exist, then hopefully we should only have to see it when we chose to do so.

I remember reading Robert Ardrey's books about anthropology. I found them a revelation, not so much because I agreed with almost every word he wrote, but because he would cut through all the crap of civilisation. He laid bare all the delusions of civilised people.

"Some animals may lie to one another, but only mankind lies to himself."

"Mankind is objective about all things except mankind himself."

I do not suggest that anything I have written in this book should be carved in stone. I hope that people will make up their own minds and decide what is right and wrong or true and false and that they trust their common sense and their own intelligence. Just because people may disagree with me on one issue does not mean that they have to disagree with me on everything.

I hope that people who read this book will be pragmatic and not regard everything I've written as 100% right or 100% wrong. Human beings evolved over millions of years and developed the ability to reason. I hope that humankind, or should I say people-kind, will never give up on reasoning, even if it appears that the rest of the world has.

THE END